3 1168 01234 5664

D0788171

An American Bestiary

Books by Jack Schaefer

Jack Schaefer

An American

Bestiary

With a Foreword by James S. Findley

ILLUSTRATED BY LINDA K. POWELL

HOUGHTON MIFFLIN COMPANY BOSTON

1975

Library of Congress Cataloging in Publication Data

Schaefer, Jack Warner, 1907–
  An American bestiary.

   1. Zoology—America.   2. Animals, Habits and behavior
of. I. Title.
QL150.S3   591.9′73   75-5630
ISBN 0-395-20710-X

Printed in the United States of America

V   10   9   8   7   6   5   4   3   2   1

". . . so we should venture on the study of every kind of animal without distaste; for each and all will reveal to us something natural and something beautiful. Absence of haphazard and conduciveness of everything to an end are to be found in Nature's works in the highest degree, and the resultant end of her generations and combinations is a form of the beautiful."

— ARISTOTLE

Contents

Illustrations

Foreword

IN TALKING about this book Jack Schaefer is quick to refer to himself, somewhat deprecatingly, as an amateur. I would be quick to agree, because to me an amateur is a lover, and it is quite clear that Jack Schaefer has a deep and affectionate involvement with his subject. Lovers of all kinds are quick, sensitive, and intuitive. They are aware of facets of the objects of their affection that are often not readily apparent to other observers, no matter how "professional." Many professionals have written on southwestern mammals, but few have noticed and commented upon the things that Schaefer takes to be noteworthy. Perhaps we professionals have taken too many things for granted, and in our eagerness to describe the new and little-known, have by-passed many obvious but uninterpreted and fascinating aspects of the lives of wild animals. It seems to me that this book helps to fill this gap and that it provides a fresh look at a diversity of phenomena with the vision and enthusiasm that only a true amateur could master.

JAMES S. FINDLEY
Department of Biology
University of New Mexico

Forewarnings

ANYONE bemused enough to wander through this book must accept the probable irritation imparted by my habit of frequently intruding myself into the text. My excuse is that what I have been recording are some of the results of my own wandering through the accumulated annals of human thought and supposed knowledge about our fellow creatures. So doing, it has seemed to me that I have been making a voyage of discovery as important to me personally, as significant to my small thinking on many subjects, as to him was Darwin's voyage aboard the *Beagle*.

For many years I was a newspaperman along the east coast of this country. Then I moved west and for even more years was a writer of short stories and novels based on Western material. Always I was writing about people, about us featherless bipeds who sum ourselves by genus and species as *Homo sapiens*. Any other creatures who crept in were merely stage furniture for the human drama. And then, as a writer, I came to a full stop. I had lost my innocence. I had become ashamed of my species and myself. I understood at last in full consequence that despite whatever dodges of motive and intent and personal activities I might cite I was a part of that deadly conquest called civilization, that simply by living and fathering children and using the products of that civilization I was a contributing

part of the heedless human onrush that was ruining the land I loved
and forcing toward extinction ever more of my fellow creatures
whose companion right to continued existence ought to be respected.

That was when I began my voyage of discovery, a pursuit of
knowledge about those fellow creatures, in particular those who are
my closest relatives, my fellow mammals. I know that the knowledge
I have garnered is spotty and full of errors and misinterpretations,
but I know too that mine are honest errors — and I console myself
by remembering that the long record of human thought about other
living creatures is riddled with errors and that to the scientists themselves today's supposed fact often becomes yesterday's mistake.

"The value of knowledge," Loren Eiseley notes in one of his
books, "is not only in what is known, it is more a change wrought by
it in the knower." A change has been wrought in me. Never again
can I regard any living thing as mere stage furniture for any human
drama. I cherish a neighborly respect and an often affectionate awe
for the abilities and the adaptations other forms of life have evolved
to solve the basic problems of existence we all must confront. I seek
and sometimes achieve a feeling not of dominion over but of fellow-
ship with all other forms of life. And bedded deep in me is the conviction that a world inhabited only by us humans and those others
we decide are useful to us would be a barren and a lonesome place.

Nearly three centuries ago the English naturalist John Ray set up
a signpost for me. "For my part," he wrote, "I cannot believe, that
all the things in the world were so made for man, that they have no
other use . . . But though in this sense it be not true, that all things
were made for man; yet thus far it is, that all the creatures in the
world may be some way or other useful to us, at least to exercise our
wits and understandings, in considering and contemplating of
them . . ."

That is my attempt, to make amends for my past attitude toward
my fellow creatures by exercising what wits and understandings I
possess in considering and contemplating of them.

In my wandering I have lingered now and again beside other signposts, directional pointers provided by writings out of the past that
summed what was known and believed about animals in their

time — and that have helped me in a general and simplistic way to mark off in my mind six periods of human knowledge in the field.

The first was the classical period in which the pertinent works of Aristotle were dominant and established the beginnings of a science of zoology based on observed data. The goal was knowledge for its own sake. Aristotle put this plainly: ". . . we proceed to treat of animals, without omitting, to the best of our ability, any member of the kingdom, however ignoble. For if some have no graces to charm the sense, yet even these, by disclosing to intellectual perception the artistic spirit that designed them, give immense pleasure to all who can trace links of causation, and are inclined to philosophy."

During this period, however, the data available remained scant. Considerable reliance had to be placed on hearsay and on garbled reports and travelers' tales and too much reliance was placed on deductive logic, on inferences as to what the reasoners thought the available data ought to mean, not what it could demonstrably be shown to mean. Then along in the first century A.D. one Gaius Plinius Secundus, usually referred to as Pliny the Elder, obligingly summed just about all of the classical zoological "knowledge" in the thirty-seven volumes of his *Natural History*.

Next came the period of fantasy and imagination, running on through what in my schooldays was invariably called the Middle Ages. The infant science of zoology faltered into decline. The few naturalists with scientific pretensions were strict authoritarians, adding little new data, settling disputes by piling up references out of the classical authorities. Meanwhile the Roman Catholic Church appropriated natural history for moralistic purposes. Books known as bestiaries became popular, surpassed only by the Bible itself. The first, the original, the *Physiologus,* appeared about 200 A.D. and on through the centuries there were scores of versions in dozens of languages and dialects. Animals real and imaginary were plucked out of Aristotle and Pliny and out of fable and folklore, given fanciful descriptions, and used allegorically to emphasize aspects of Christian dogma. Frequent additions were made until the original tally of 49 animals had risen to about 160. Saint Augustine himself stated that it was of no consequence whether these animals actually existed; what mattered was what they signified.

In a scientific sense the bestiaries were downright silly, but one aspect of them seems to me to deserve respect. By their very concept they recognized, they accepted, that man has within him something of a "beastly" nature, is to that extent one among the animals. They did not divorce man from the rest of the natural world, as later did the so-called Romantic Movement whose effects still cloud much modern thinking.

During the next period, roughly the sixteenth and early seventeenth centuries, men like Konrad Gesner of Germany were reviving zoology as something of a science. They still looked backward, spent much of their time assembling every reference to every animal they could find in all past writings to which they had access. But they questioned the wilder and more fanciful creatures and notions, and they added a fair amount of new material, including data beginning to be brought back by the early explorers of new lands across the seas. They were handicapped by lack of adequate terminology and a practicable system of classification. But they were re-establishing zoology as the study of real animals, of those believed to be real, and again now primarily for the simple sake of knowledge about them — though the purpose usually professed was to praise the glory of God by revealing the wonders of His creation.

The most important work of the period was probably Gesner's four-volume *Historia Animalium*. Fortunately for those of limited scholarship like myself, this became the basis of the Reverend Edward Topsell's *The History of Four-footed Beasts,* published in London in 1607. This in turn was, I believe, the first natural history in the English language. It was reissued in 1658 with two thinner volumes covering *Serpents and Insects.* Though Topsell noted on his elaborate title pages that his material was "taken principally" from Gesner, he supplied additions out of his own research and contemporary gleaning of data. The whole is a big fascinating folio compendium of just about everything known and believed about our fellow creatures at the time.

My next period, primarily the eighteenth century, is the one in which the science of zoology really began to take shape. The most significant contributor was Carolus Linnaeus of Sweden. The ten editions (1735–1758), each expanded and improved, of his *Systema*

Naturae launched at last the systematic classification of living creatures with which later zoologists have been making ever more sense out of the vast and intricate puzzle of life on this earth.

During the same years Georges Louis Leclerc of France, who became the Comte de Buffon, was compiling his all-inclusive *Histoire Naturelle.* The first part was published in 1749 and by the time of his death in 1788 all of thirty-five volumes had been issued and thirty-six more were in preparation. And meanwhile over in England none other than Oliver Goldsmith was repeating with Buffon's material what Topsell had done earlier with Gesner's. His eight-volume *An History of the Earth and Animated Nature,* published in 1774, was based chiefly on Buffon's work to that date with additions of his own notions and those current in England at the time.

I find in Goldsmith's work following Buffon the two currents of thought, of attitude toward natural history that have since gone ever more divisive ways. They are not confined to natural history but run through all the sciences. Though they are often mingled in varying degree in an individual scientist's mind, as with Goldsmith, the distinction between them is of considerable significance.

One goes back to Aristotle and is characteristic of what I call "pure" scientists, those motivated primarily by human curiosity, by the desire simply to know, to expand the frontiers of knowledge for its own sake. As Goldsmith saw this: "A man of this disposition turns all nature into a magnificent theatre, replete with objects of wonder and surprize . . ." Such a pursuit, he claimed, "gives higher satisfactions than what even the senses can provide." The other traces back to the "dominion" thesis of the Old Testament, the assertion that man was specially created to have dominion over the earth and all things in it. This attitude is reflected in the efforts of "practical" scientists, those who pursue knowledge primarily for the uses to which it can be put, the application of it to practical human purposes. Goldsmith phrased this in terms of his belief that the natural world is no fit habitation for man, is "but a desert place, without human cultivation," and that God created man to improve it. "God beholds with pleasure, that being which he has made, converting the wretchedness of his natural situation into a theatre of triumph; bringing all the headlong tribes of nature into subjection to his will;

and producing that order and uniformity upon earth, of which His own heavenly fabric is so bright an example."

That latter attitude has supplied much of the motive power for much of what I have already called "the heedless human onrush" that nowadays is overtaxing the world's resources and threatening the existence of so many of our fellow creatures. New England's Cotton Mather was, I believe, the one who carried it to its blunt inevitable extreme by asserting that what was not useful to man was vicious.

In contrast I recall a statement by Thoreau reported by Emerson: "Thank God, they cannot cut down the clouds." And yet . . . that is precisely what the rainmakers and weather manipulators are currently trying to learn to do.

In my next period, roughly a century, the 19th, the science of zoology came of age and the conceptual foundation was laid for all later work in the field. As a period it offers a variety of signposts, but two stand higher than the rest.

The first was erected by Jean Baptiste Pierre Antoine de Monet, Chevalier de Lamarck. Much of his work was done in the latter part of the eighteenth century, but his major mature writings were not published until the nineteenth was under way. His contemporary, Baron Georges Cuvier, set the fashion still followed by many people of sneering at Lamarck's theory of acquired characteristics, the notion that a new structural adaptation in an animal species can develop from a "need" (Cuvier twisted this into a "wish" in his criticism) and thereafter such a change is transmitted by heredity. That is, the adaptive responses a living creature makes to its environment can cause structural changes that are inherited by its progeny. An example often cited to illustrate the theory is that the giraffe, needing a longer neck to forage in trees, kept stretching what neck he had and thus elongating it and each generation passed the acquired elongation on to the next in successive stages.

Lamarck was wrong — but he was also very right. He was wrong in the means, the mechanism, he offered for structural changes — but he was right on the much more important point that changes have occurred and do occur. At the time almost all other zoologists (Cuvier a leader among them) were still convinced of the fixidity of

species, of the eternally ordained original creation of them. Lamarck was introducing the concept of "transformism," of the modification of species and the development of new species, which before
long would be known as "evolution."

The man who erected the other tall signpost, somewhat later in
the century, was, of course, Charles Darwin. His stands very tall
because of the thoroughness of his work, because of his discovery of
"natural selection" as the mechanism of adaptive changes, and because by about the end of the century those who championed him
and his work had won the battle, the fundamentalists and special
creationists were reduced to making feeble skirmishes in retreat, and
the theory of biological evolution, with man as one though a very important one, of its products, had become the basic concept underlying the science of zoology.

My sixth and last period is open ended, covering this century to
date and continuing. Zoology has become a vast field of study in-
timately connected with and drawing data from an increasing
number of other biological sciences and from the physical sciences as
well. It has become an arena for an increasing array of specializations with many thousands of specialists doing research and publishing books and papers. The extent of their specialization often
amazes me. In an extreme case, for example, a one of them may
spend several years studying one aspect of one parasite of one internal organ of one animal species — and do this unaware that several
others in other places are pursuing almost the identical project. In
all fairness, I should add that some of them are trying to develop
better coordination of ongoing research.

In all probability more zoological data of one kind and another is
nowadays accumulated in one year than in all the years of human
history up to the time of Darwin. No one, not even a super-Aris-
totle, could keep up with it. Individual zoologists are hard pressed
just to keep abreast of what is happening in their own areas of
specialization, and find it increasingly difficult to achieve a wider
view. But in general all that is being done is held together by two
basic concepts, two assumptions accepted throughout the field. One
is that evolution provides the plot line for the story of life on this
earth in all its manifestations. The other, that man is a natural

being, a part of the natural world, one of the late-arriving characters in that story.

It seems to me that nowadays the gap between professional knowledge and lay knowledge is greater than it has ever been. Not just with regard to the multitudinous data but also with regard to the implications drawn from them. At the very time biological scientists are piling up amazing amounts of evidence, directly or indirectly supporting the validity of those two basic concepts, outside scientific circles all the old notions of fixed creation, of man as something apart from nature, of the natural world as divinely ordained for his exclusive use as he wills, are still very much with us.

It seems to me, too, that the science of zoology is becoming so "scientific," so devoted to the accumulation of knowledge whether for "pure" or "practical" purposes, and so many of its devotees are becoming so objective in attitude toward their work, so single-minded in pursuit of fractions of specialized data, that zoology is becoming less a study of living creatures and more a study of objects considered of value solely as subjects for study — and for the acquisition of reputations and scholastic degrees. I know that the piling up of verifiable facts is one of the major requisites of scientific advance. But the current concentration upon it by such an expanding army of researchers both "pure" and "practical" seems to me to have somewhat the flavor of a return to the notion of our fellow creatures as existing solely to be of use to man, to provide him with data to be piled up, with facts to be preserved in print.

I note that more and more of the specialized research under way requires the capture and killing and dissection of more and more of the creatures (man excluded) being studied. When I add this to the accelerating pressures our human population explosion is putting on the very existence of so many species, I sometimes conjure up a daytime nightmare to the effect that the time will come when the busy researchers will have piled up every conceivable verifiable fact about our fellow creatures — just as the last of them has died. We will then have all possible knowledge about them, but we will not have them sharing the experience of life with us.

Robinson Jeffers knew what ails us humans:

> A little too abstract, a little too wise,
> It is time for us to kiss the earth again.

In a more optimistic mood I sometimes think that perhaps a new period, a seventh, has begun. I base that on the emergence of a relatively new branch of zoology known as ethology, the study of animal behavior. As understood in the modern sense this had its beginnings early in this century, but only quite recently has its importance achieved wide recognition.

From the beginning, of course, zoologists have paid some attention to the behavior of animals, even during those periods in which major emphasis was on efforts at the systematic classification of them, which of necessity was (and is) concerned chiefly not with behavioral but with structural and physiological data. The point is, however, that until lately almost invariably behavioral accounts were based on very limited and highly distorted evidence.

For a long time all kinds of misinformation came from fable and folklore. As much or more came from reports of explorers and travelers, most of whom, being humans, usually regarded any animals encountered as their rightful prey to be killed for use as food or trophies or scientific specimens to be taken back home. In such encounters man was almost always the aggressor and such reports of animal behavior were of behavior limited to response to human aggression. By what psychologists call transference the reporters transferred to the animals some of the characteristics they themselves were displaying. That transference habit is still frequently reflected in magazines catering to hunters. The animals hunted, refusing to be willing victims and sometimes even fighting back, become sly and cunning or ferocious and bloodthirsty creatures, while the hunters are the ones displaying admirable skills and praiseworthy courage and performing the deeds of derring-do.

Even when the scientific approach became dominant, the behavioral data remained poor. Not much (and that only by inference) can be learned about the actual behavior of an animal species by dissection and study of carcasses. Not much more (and that chiefly of abnormal or subnormal behavior) can be learned from observation of animals penned in zoos or experiments with animals confined in laboratories. Yet those were the usual methods employed.

The core of ethology today is the study of the behavior of animals in their natural habitats. It is the study of living creatures leading natural lives. It is as exacting as any other scientific discipline in its need for accurate data, demanding, as Konrad Lorenz has emphasized, the attempt to become acquainted with "the *entirety* of the observable behavior" of the species studied. By its inherent rules it encourages a minimum of disturbance of the subjects studied — and no killing. I would call it a friendly, a neighborly pursuit of knowledge, one that leads to increased understanding and appreciation of our fellow animals as *living* creatures.

To my mind the tallest ethological signposts to date are being erected by Lorenz and by Niko Tinbergen. With the exception of rare little volumes like Lorenz's *King Solomon's Ring* and Tinbergen's *Curious Naturalists* most of their published work offers hard going for an amateur like me. More readable are such recent definitely ethological books about individual species as Farley Mowat's *Never Cry Wolf*, Victor B. Scheffer's *The Year of the Whale*, Jane van Lawick-Goodall's account of the chimpanzee *In the Shadow of Man*, and G. B. Schaller's *The Year of the Gorilla*. Such, and their number increases, are really "popularizations" of the actual work done, which is available elsewhere in more scientific form. But they sum those results in fascinating manner and speak honestly to the lay mind — and can jolt it into realizing that all too much of what we humans have all too long believed about the characters and behaviors of our fellow creatures has been very much distorted or even flatly wrong.

In time ethology may even teach us that we have not come as far along the evolutionary road as we like to think we have, that we still have more evolving to do — that Lorenz is right when he states that "the long sought missing link between animals and the really humane being is ourselves."

With women's rights movements so active nowadays, perhaps I should apologize for my almost invariable use of masculine pronouns in reference to the mammals I discuss. Our language is lamentably lacking in singular pronouns that could be understood to refer indiscriminately to either sex. "It" and "its" are the only ones available — and I refuse to refer to a fellow mammal as an "it." Stuck with he-or-she and him-or-her and his-or-hers, I have allowed my own gender to predominate in my usage.

I have also allowed my own current habitat, the Southwest in general and New Mexico in particular, and its wildlife inhabitants to predominate in my discussions: no apologies for that and with good reason.

New Mexico, Land of Enchantment. We New Mexicans like that label. We constantly flaunt it — embossed on the license plates of our motor vehicles. But is it true? Of course it is.

I read an essay by Aldo Leopold and suddenly am startled. Unashamedly, openly in print, he refers to the Southwest in general and New Mexico in particular as "this wrecked landscape." Is that true? Of course it is.

Both labels, one emotional, the other ecological, are correct. They can be reconciled thus: New Mexico, despite the wrecking of landscape, despite the deterioration of soils and plant and animal life wrought by man, is still a land of enchantment. It has been tough enough, inhospitable to so-called progress enough, had natural resources enough, to retain a fair portion of its enchantment.

Nature herself must be thanked. Not alone for being sparing of rainfall, thereby slowing the growth of contemporary civilization here, but also in providing variety of environment and of animal life.

The United States as a whole can claim seven life zones. Most states have two or three or four of these. New Mexico has six of the seven, ranging from the Lower Sonoran all the way up to the Arctic-Alpine. A corollary of that range of life zones is a corresponding range of wildlife.

Consider mammals. Leave out the primates, that topheavy order almost all of whose members prefer life in other continents than North America. That leaves twelve orders, major groupings, of mammals listed by taxonomists as belonging to this continent. One is found only in tropical Mexico and Central America. Three cover mammals of the oceans and seas. That leaves eight for the land mass and lakes and rivers of this continent north of the tropics, including of course all of the United States except Hawaii. New Mexico has representatives of all eight of those orders. It has representatives of almost all major families of mammals within those orders, of most of the major genera within those families.

I leaf through Vernon Bailey's *Mammals of New Mexico,* North American Fauna No. 53. Some of his species have become subspecies, some of his species names have been changed, various new species of bat have been identified since his time, and it really would be a problem to attempt a cross-comparison of his rodents, most numerous of mammals, with later classifications. Nit picking. His work stands up remarkably well. And suddenly I am startled again. He lists for New Mexico six species of grizzly bear — and even adds a subspecies. That was in 1931. Not long ago I was looking at a grizzly hide displayed in one of our state buildings, that of the last wild grizzly seen alive in the state. Killed in 1933.

Bailey was certainly looking backward in his listings. He knew he was. He knew that at the time he was writing the jaguar and the bison and Merriam's elk were already extinct here. The Rocky Mountain bighorn and the Mexican mountain sheep and the northern elk were "almost if not entirely destroyed." The pronghorn was "rapidly shrinking in numbers." Deer were having "their range and abundance" reduced and the Sandhill white-tailed deer was probably gone. The peccary had "only a slight foothold" left.

Wrecking crews had worked on more than the landscape.

Most of the species mentioned above are in better circumstances now, some restored to the state by importation of breeding stock. Most are protected by game laws. That word "protected" is somewhat limited in meaning. It denotes rules and regulations designed to ensure that in the annual "harvesting" of game by hunters enough animals will be left alive to continue the various species. That is: the species are protected, not the individuals as such within them. I find irony and a reflection on the rest of us that much of the support and most of the financing for this "conservation" of wildlife, here as elsewhere, has regularly come from people devoted to the killing of wildlife. Hunters and fishermen. Hunting and fishing fees. A fascinating motive behind this for psychologists to study: preserve a species so that there will always be some to be killed. Selective about that too. Withhold protection from other predators so that their numbers will be reduced toward extinction and thus you will have an excuse, a rationale, for your own killing of what would have been their prey.

Two thorough surveys have been made of New Mexico's so-called game resources. The first was prepared for the then Game Commission back in the mid-1920s by J. Stokley Ligon, a Federal Government specialist. About time. As he noted in his introduction, a "crisis in New Mexico game conditions" had been reached by 1924. "All species were at their lowest ebb of production and were most threatened about this time." His full report, *Wild Life in New Mexico,* published in 1927, remains an interesting document.

The second survey, published forty years later in 1967 under the title *New Mexico Wildlife Management,* was prepared by staff members of the Department of Game and Fish. It too is an interesting document. Set side by side the two are doubly interesting.

Specialist Ligon fully deserved the tribute paid him by the second survey, which was dedicated to him as one who had devoted "his life to wildlife conservation." And yet, leafing through his report, I come on this: two photographs, one of the corpses of a bobcat and a coyote killed by a government hunter, the other of a coyote struggling to escape from a trap. Underneath these: "Wildcats and Coyotes are the eternal enemies of wild life."

Strange — I always thought that wildcats and coyotes were wildlife too. I think I prefer the definition of wildlife enemies given by cartoonist "Ding" Darling, who was quite a conservationist in his time: "The worst enemies of wildlife are the Republicans and the Democrats."

I cannot blame specialist Ligon. He simply shared the prejudice of his period. Here, for example, is a volume of 1931 titled *Thirty Years War for Wild Life* by zoologist William T. Hornaday, virtually the generalissimo of all conservation groups in Ligon's time. What was Hornaday's attitude toward creatures like wildcats and wolves and coyotes? "The coyote should always be killed." "The gray wolf should always be killed."

Simple enough to sum the Ligon-Hornaday attitude: any creature who offered competition to man in the hunting of game was an enemy of wildlife.

Time passes and attitudes change. The 1967 survey offers more pleasant reading — and not just because it can cite improvements in game management, species saved, others restored by stocking, etc. The whole tone is subtly different. Particularly in the section on "predators," which points out that man is definitely a predator himself and notes that there are "different opinions" about the others and is content to discuss them as interesting fellow creatures. I suspect that Louis Berghofer, who wrote the section, has a fondness for them. He wants the reader to know that the puma has "contributed in no small part to the folklore of this state," that the "outlook for the bobcat appears good," and that the howl of the coyote will "continue to be heard in New Mexico for many years to come."

That shift in attitude, I believe, is beginning to produce results. I note with some pride what I consider advances here in my New Mexico. That competing predator, the puma, has recently been given protection by being declared a game animal and thus the hunting of him is now subject to season and permit limitations. And just this year, 1974, a new division has been created within the Department of Game and Fish to work in behalf of all endangered species — and to finance the new division money has been appropriated for the first time to the Department out of the state's general fund. It is a small beginning, not much of an appropriation, but all of us tax-paying New Mexicans are now sharing in the good work.

The other day, just for the hell of it, I started calculating what might be represented in a hunting drama, a single hunter out after a single New Mexican deer. Both are culminations of immensely long processes of physiological evolution tracing back to the emergence of the mammalian class and on back to the beginning of life on the earth. In that respect they start on an equal basis. But that is all the deer has, plus the small amount of "education" acquired from a few other deer and his own experience to date. The hunter is armed with the results of another long and very important process, that of cultural evolution. He is being helped in his hunting by a huge army.

He gets to the deer's habitat in a jeep. The jeep alone, in terms of the experiments and inventions behind it and of materials and manufacture and delivery, sums the efforts of thousands upon thousands of people. It is almost the same for the clothes he wears. The same again for the rifle he carries and its telescopic sight and the ammunition he puts into it. He has in his mind information from a multitude of people, not to mention those who made possible the printing of that information in books and magazines he has read. Without trying very hard I ran up a total of millions of helpers.

Taken there by a jeep created by others on highways and roads prepared by others, protected against weather and brush by clothes made by others, guided by experience and knowledge passed on to him by others, he meets or stumbles upon the deer. He raises the rifle designed and produced and put into his hands by others and pulls the trigger and a bullet propelled by a smokeless powder whose chemical composition he may or may not know kills that deer. He thinks he killed it.

He did not kill that deer. Mankind killed it.

Something is missing from my brain, a convolution or circuit, the one that would tingle with a thrill were I that hunter and I shot that deer. Or any other fellow animal, certain insects and arthropods and that foreign invader the house mouse excepted. I am not against killing in principle. I enjoy a good steak. I have chopped off the heads of dozens of chickens and a few turkeys. But the deliberate killing of a wild animal except in a case of emergency would give not a thrill but a pang. A sense of loss.

My experience in the deliberate killing of wild animals has been confined to three incidents. One occurred when I was a boy in Ohio and a rat was regularly reducing the number of young rabbits I was trying to raise. After various misses from various ambushes, I got him: no thrill even though he was another foreign invader, a so-called Norway rat. Just a job done and burial given. The second was many years later at our ranch out Cerrillos way here in New Mexico. One of our girls was making a daily jaunt to a neighbor's place to feed two cats and a turtle while the neighbor was away. An excited call came on the old crank phone that a rattlesnake was in the walled patio there. I grabbed a .22 pistol and hurried over.

Right enough. A snake was there, slithering through low vegetation, rattlesnake markings on what could be seen of him. I aimed and fired.

I am not like a fellow New Mexican writer who asserts, in print too, that when he shoots at a "critter" "it is supposed to drop . . . as if struck by lightning." In my case, bothered by a buzzing fly would be more apt. But this was my unlucky day. I got that snake first shot. Then, inspecting the corpse, I discovered it was not a rattlesnake. It was a bull snake. I learned later from the neighbor that it had been virtually a pet around his place.

The third incident involved another snake: a real rattler this time. It came crawling out from under the small stoop on the chickenhouse at the ranch. Thinking only of grandchildren and chicks about, I picked up a shovel and smacked off his head. No thrill. On second thought I decided I should have caught him, taken him into the back country, and let him go to enjoy many a meal on assorted members of the ubiquitous twenty-two species of New Mexican mice.

John Steinbeck learned "the best of all ways to go hunting." He learned it while with a hunting party after *borregos,* bighorn sheep, in Mexico. They found no sheep. They did find some dried sheep droppings. He had one mounted on a small hardwood plaque and thereafter could point to it and say: "There was an animal and for all we know there still is and here is proof of it. He was very healthy when we last heard of him."

I am aware that my discussions of various fellow mammals in this book can be condemned as Lamarckian — worse yet, as following Cuvier's twisted version of what Lamarck really meant. That is: I write about them as if in evolving adaptations to their environments they were responding not only to needs but to wishes. I write about them as if they have *wanted,* have *willed,* to evolve in this direction or that. I have done so deliberately. It is a convenient device for tracing their separate routes through the evolutionary maze to the present.

I am aware too that often I indulge in another fault that can affront the scientifically orthodox. I sometimes anthropomorphize, discuss animals and their habits as if they can be evaluated ethically in human terms, refer to them as if they had human motivations. Again I have done so deliberately. It is a convenient device for describing (not necessarily explaining) animal behavior in commonly understandable terms.

Both devices are quite unscientific. But I have not been writing (am not qualified to write) a scientific treatise. I have merely been considering and contemplating of some of my fellow creatures on the basis of some personal experience and much reading in the field and a mild amateurish smattering of scientific knowledge. I claim for myself only as much as does the character in one of Shakespeare's plays who states: "In nature's infinite book of secrecy, a little I can read."

JACK SCHAEFER

Los Ranchos de Albuquerque

An American Bestiary

Order: *Artiodactyla*
Family: *Tayassuidae*

The New World Pig

I REST my right hand on my desk with the weight on all five fingertips pressing down and evenly spaced apart. I have there a symbolic representation of the five-digited feet of my remotest mammalian ancestors.

I raise my arm higher and higher while maintaining the finger pressure and soon the first and fifth digits (my thumb and little finger) have lost contact with the desk and the pressure is sustained by the middle three. I now have a representation of the feet of certain early mammals who left the evolutionary road toward me to go in another direction, pushing themselves toward a tiptoe posture presumably in a reach for greater running speed. Some of them were satisfied with this three-toed stance and their surviving descendants still use it.

I continue raising my arm and soon the second and fourth digits also have lost contact and the pressure is solely on the middle one. I now have the final tiptoe stance achieved by my friend the horse and his relatives. In a few seconds I have gone through the five-three-one progression worked out through millions of years by the hoofed mammals with odd-numbered toes.

I start again with my hand placed as before. If I rock my arm and adjoining hand slightly to the right, the first digit (my thumb) loses

contact and the pressure is sustained by the other four fingers. I
have a symbolic representation of the feet of certain other mammals
who also left the evolutionary road toward me to go in yet another
direction. Some of them were satisfied with this four-toed stance
and their surviving descendants still use it.

I raise my arm higher and soon the two outside of the four digits
have lost contact and the pressure is sustained by the middle two. I
now have the final stance of my friend the deer and his relatives. In
a few seconds I have gone through the five-four-two progression
worked out through millions of years by the hoofed mammals with
even-numbered toes. And at this moment I repeat to myself my
vow never to refer to the two-toed as "cloven-hoofed" despite the
Biblical and other respectable uses of the term. Their feet are not
cloven-hoofed. No cleaving ever took place. Each foot has two
hoofs developed from two digits side by side.

I find it interesting though time-taking to try to trace the in-
tricacies of the scientific classifiers, the taxonomists, who are con-
tinually revising their catalogues of creatures as knowledge of these
increases — and sometimes as their personal prejudices dictate. Debate is usually between the splitters and the lumpers, those who like
to split their subjects into more and those who like to lump them
into fewer orders and families and genera and species. In the
search for differences in the one direction and relationships in the
other, they even resort to terms like infraorder and suborder and
superfamily. Confusing at times, their work is yet invaluable to an
amateur like me, trying to find his way with some semblance of direction through the wondrous maze of life forms inhabiting the
same world with him.

For a time the taxonomists lumped all hoofed mammals into the
one order Ungulata. Nowadays the splitters have prevailed and two
orders are generally accepted, Perissodactyla for the odd-toed, Artiodactyla for the even-toed.

These two groups that have gone their own foot-ways must have
started with about equal chances. Yet one group, the artiodactyls,
has been much more successful.

According to the fossil record the perissodactyls, the odd-toed, have tried hard to hold an important place in the evolutionary parade, achieving through the ages quite a list of families and genera and species. Unfortunately for them, however, they have also achieved the highest extinction record of any of the surviving mammalian orders. They are represented nowadays by just three small families with only 5 genera among them and 16 species — and most of those species (the horse with human aid is the major exception) have very limited ranges and some are currently close to extinction. The artiodactyls, on the other hand, have achieved a wider radiation into varying life forms and have had a much lower extinction record. They are represented nowadays by 9 famililies, at least 80 genera, and some 194 species of which many have extensive ranges and large populations; all together they are virtually worldwide in distribution.

Certainly the artios have not been more successful than the perissos because even-numbered toes-become-hoofs are more efficient for foot purposes than are odd-numbered. They have done better, I believe, primarily because some of them went ahead to develop another specialty. These became ruminants. They enlarged their stomachs and divided them into compartments, usually four. Thus equipped they could shorten their mealtimes, usually dangerous times because of the difficulty of simultaneously garnering a dinner and keeping watch for approaching predators. They could cram down food unchewed into the first compartment, then retire to some safer place to enjoy it at leisure. Meantime that food would be moving into the second compartment where it would be softened and squeezed into small batches we humans call cuds. They could then regurgitate these cuds for unhurried chewing and mixing with saliva before consigning them to the third compartment from which they would move into the fourth, absorbing gastric juices along the way. By time the food passed into the intestines it would be thoroughly prepared for digestion of whatever nutrients it contained. Added advantage of such a digestive specialty was that those artios who took to ruminating could expand their menus, could find some food value in almost anything in the vegetative line.

To me there is conclusive proof that this specialty favored the artios in the fact that the ruminants comprise about 90 percent of the entire artio roster. Drop them out of the reckoning and the remaining artios just about match the perissos.

That odd-even division is a neat device for taxonomic purposes. But it is not just a matter of counting toes. Consider the tapir. He has four toes on his front feet, three on his hind feet. Should he be artio or perisso?

The division works only when its application is confined to those toes which are actual weight-supporters. Some of both the perissos and the artios still have toes which they have not bothered to evolve away even though these no longer really function as toes. For classification such toes do not count in the counting. To avoid possible arguments the taxonomists usually describe the division in non-toe terms. What they call the distinguishing diagnostic feature is the axis of the foot, an imaginary straight line running forward and bisecting the functional area of the foot. For the perissos, the odds, the axis leads straight to the original middle digit — the one that has become the single hoof of the horse. For the artios, the evens, the axis leads to a point midway between the original third and fourth digits — the two which have become the twin hoofs of the deer.

So the tapir is perisso, not only because three toes are the weight-supporters, but more scientifically because the axis of each foot leads straight to the original middle toe. The extra toe on each front foot is a dew-claw, an extra no longer used as a toe, which he has not got around to discarding.

The peccary also has four toes on his front feet, three on his hind feet. But he is an artio because only two toes on each foot are weight supporters and the axis aims at a point midway between the original third and fourth digits.

There. At last and by roundabout way I have arrived at a first mention of the fellow mammal of whom I am currently considering and contemplating. The peccary. The New World pig.

In simple justice it is no more accurate to refer to the peccary as the New World pig than it would be to refer to the pig as the Old World peccary. The pig has advantage in the naming not because of any evolutionary priority but solely because Europeans were well acquainted with him long before they knew the peccary even existed. When they did meet the latter, seeing a definite resemblance, they applied pig labels. Early explorers in the New World called him bush pig or musk hog.

Undoubtedly the two, pig and peccary, had a remote common ancestry. But they have gone their separate ways for more than thirty million years and represent a case of somewhat parallel development rather than any close kinship. They resemble each other primarily in that both have chunky block-bodies, legs slender in proportion to body bulk, twin hoofs on each foot, and longish snouts ending in flat disklike tips. To me, however, the differences are more interesting than the resemblances.

The pig seems to have been content very early to go only a short way along the artiodactyl route. He is probably the most primitive in this sense of all the artios. He still has four toes on each foot — though he has reduced the side ones to some extent, making the middle two the main weight-supporters. He has remained short as well as slender legged and in most forms he is a relatively slow runner. Moreover, he apparently has tried few if any experiments with his stomach.

The peccary went further along the artiodactyl road before being satisfied he was in shape for survival. Starting from the same four-toed stance, he placed greater emphasis on running, went less for sheer body bulk, and developed relatively longer legs. As time went on he fused two median bones of each foot into the characteristic ruminant cannon bone and relied more and more on the two-toed stance, eliminating altogether one of the now-extras on each hind foot, reducing the other and the two now-extras on each front foot to mere remnants. In addition he began to complicate his stomach with definite suggestions of compartments. While he stopped short of becoming a true ruminant, he certainly headed in that direction before deciding he had gone far enough.

The pig in all his forms has a tail. This is nothing for him to boast about, but it is certainly a tail. The peccary has none at all, and has only a slight protuberance on his rear where a tail would be. But he has something the pig does not, a scent gland in a skin fold on his back not far forward from where his tail would begin if he had a tail.

As far as I know, he is the only mammal who sports a scent gland in such a position. The musky odor it can emit is similar to but not as strong as that of the skunk — though even a weak human nose can detect it at quite a distance. The suggestion is sometimes made that he has developed this as an insect repellant to take the place of the switching tail he lacks. I doubt that. While he apparently can also activate the gland at will, it invariably goes into automatic action whenever he is startled or angry. The odor emitted may have some effect as a repellant to some intruders, but its major purposes are undoubtedly to serve as a warning signal to his fellows in time of danger and as a means of communication with them as suggested by his habit of rubbing his back against bushes and other objects to leave traces of his presence in the neighborhood.

He has resolutely refused to parallel the pig in another important matter. Particularly in masculine incarnation the pig has turned his canine teeth into varieties of tusks whose major characteristics are length and curvature. The upper sets do strange things. Instead of growing downward like sensible teeth, they grow outward and curve upward and back. The climax in this kind of nonsense has been attained by the pig known as the babirusa, who inhabits some of the East India Islands and whose upper tusks have sockets switched upside down so that those tusks grow right up through the upper jaw and out and continue up and curl back until they almost poke him in the eyes. With the lower ones also sweeping out and up and curving back, it is a wonder he can see any way but sideways. As teeth all four tusks are useless; as weapons they are virtually useless. They have probably become mere symbols of sex and social status.

The peccary has not indulged in any such dental displays. He does have well developed canine teeth, sturdy and sharp and rather long. The upper grow straight down, shaped like spear points. The lower grow almost straight up, curved a bit on the forward edge. The two pairs interlock as his jaws close, the upper sliding down just behind the lower and the tips of those lower fitting into pockets in his upper jaw. When his mouth is closed, they are all out of sight — inside his mouth where they belong. The four are what canine teeth were originally designed to be, biting and tearing teeth. As an aid in their use his jaw gape is proportionately wider than that of the pig, who in some of his forms, especially the wild boar, can do effective slashing with his canines-become-tusks but is a poor biter. The peccary can do a fair share of slashing with his canines that are still canines and is also an efficient biter.

In my opinion the male peccary is more of a gentleman than is the male chauvinist pig. While the male pig is always larger than the female, the male peccary is content to be the same size as his mate. While the male pig has decked himself with tusky ornaments well beyond those of the female, the peccary has permitted his mate to match him in dental equipment. While the male pig likes to monopolize females and have a harem available, the male peccary is satisfied with a one-to-one ratio in his populations. While the male pig is not exactly companionable toward his females except when interested in sex and then is rather rude about it all, the male peccary keeps friendly company with his mate almost all of the time. While the male pig expects his females to work hard at motherhood, producing an average of eight or more young at a time and frequently several litters a year, the male peccary is content with a mate who produces only two young at a time and only one batch a year.

Though I have been citing differences, the fact remains that the pig and the peccary resemble each other more than either resembles any third animal. Though neither has any really close relative, each is the closest the other does have and inevitably they are grouped together in taxonomy. Just as inevitably, since the pig was earlier known to taxonomists, they are grouped together as pigs.

They belong, of course, to the order Artiodactyla, which includes the pigs, the hippopotamuses, the camels, and the multitudinous ruminants. The last two groups have their own separate suborders, leaving the pigs and the hippos lumped together in another suborder labeled Suiformes, which my memory of long ago Greek and Latin classes translates as "pig-shaped." Then the hippos have an infraorder of their own, leaving the pigs in their infraorder Suoidea — simply "the pigs." Then at last comes the separation into two families: Suidae for the Old World pigs; Tayassuidae for the New World pig — or, as most taxonomists are willing to call him, the peccary.

I have tucked a small but important distinction into the preceding

sentence: Old World pigs, plural; New World pig-peccary, singular.
Nowadays there are five genera of pigs, one genus of peccary. That
is: the true pig has five different though closely related surviving life
forms ranging from the pygmy hog through the wild boar (from
which most domestic breeds have been developed) to such tusk fanciers as the wart hog and the babirusa, while the peccary has just the
one single surviving form.

Again, I have tucked the word "surviving" into the preceding sentence because both pig and peccary have tried adaptive experiments
in the past that may have been successful in their times but are no
longer. The pig has had four subfamilies with a variety of genera
now all extinct and the present surviving line has lost some eight
genera. The peccary has had at least one subfamily now extinct and
his present line has lost some seven genera.

That dwindling of life forms, paralleled in varying degrees
through all the mammalian orders, is in part a "normal" aspect of
the evolutionary experimental advance, but the extent to which it
has occurred has ominous overtones. Some evolutionists suggest
that the current so-called age of mammals is on the wane. The
dwindling started before we humans made our arrival on the world
scene and for quite a while we played little or no part in it, but ever
since we shifted from being hunter-gatherers to being agriculturists
we have been helping it along and at an accelerating pace. Nowadays the list of endangered species steadily increases. Quite possibly
we humans represent the culmination, the final phase, of a vast experiment with the mammalian mode that Nature in due time will
cancel out or, as she did with the reptilian, reduce to relative unim-
portance to make way for another.

Carolus Linnaeus, the early Swedish taxonomist who established
the binomial labeling system (genus plus species) now universally
used, considered the peccary just another pig and lumped him into
the major pig genus *Sus*. Before long, however, the peccary's right
to at least his own genus was recognized and on through the years
various labels were offered. The most ingenious was supplied by
the French naturalist Georges Cuvier. Bemused into the belief that
the peccary's scent-gland skin-fold was a second navel, Baron Cuvier
concocted *Dicotyles* from the Greek for "double" and "cup-shaped

hollow." That one enjoyed quite a vogue on into this century. Nowadays there is general agreement to use one of the early genus names, *Tayassu*, and when his additional right to his own family was also recognized, the family name Tayassuidae was based on this.

Agreement is not as general on names for his two species. All the authorities I have checked now accept *T. tajacu* for the collared peccary, but they are about evenly divided in regard to the white-lipped peccary. Some use *T. pecari*, while others, devoted to Latinizing, insist on *T. albirostris*.

Pecari was taken, I believe, from an Amerindian name for him and is the basis for what I insist is his correct common name. I suspect that *tajacu* had a similar origin — and I note that nearly two centuries ago Goldsmith's description of him following Buffon was headed: "The PECCARY, or TAJACU."

With two such appropriate and now time-honored names available, I can attribute only to the southwestern Spanish-Mexican influence the peculiar insistence of the Fish and Game Department of my New Mexico upon calling him the javelina. They are labeling him a feminine Old World pig. Javelina is derived from the feminine of the Spanish for the European Wild Boar: *jabalí,* male; *jabalina,* female.

Already well started on their separate evolutionary paths, the pig and the peccary entered the known fossil record in the early Oligocene at least thirty-five million years ago. At that time the peccary was the more far ranging. Both were Europeans — but the peccary was also a North American. By the late Miocene, say fifteen million years ago, the pig was the wide ranger, having expanded his coverage on into Asia, while the peccary had become extinct in Europe and was now solely North American.

I suspect that for a long time thereafter the peccary had a relatively restricted range confined primarily to the southernmost portions of this continent. He has no underfur (which the wild boar does have) and the coarse bristly hair coat he wears gives scant protection against cold. But when the Central American land bridge was restored, say two million years ago, he began colonizing southward into South America and found vast areas to his liking. By the time Columbus sailed the ocean blue he had an overall range extending from the southwest of what is now the United States all the way southward to mid-Argentina.

His two contemporary species have divided (with some overlapping) the North and South American regions whose climate suits them chiefly on the basis of habitat preference. The collared peccary, so named for the yellowish-to-whitish stripe circling his shoulders, prefers relatively open country and semiarid, even arid, areas. The white-lipped peccary, whose yellowish-to-whitish coloration covers virtually his whole snout, averages the larger by a mild margin, is somewhat darker in body color, and has a fondness for forested areas.

With such a fondness, particularly for tropical forests, the white-lipped is content to venture no further north than southern Mexico. It is the collared who ranges on up into the Southwest and thus has representatives in my New Mexico.

Here he is, then, the collared peccary, the only one of his kind still at home in this part of the continent his ancestors made their homeland long ages before the ancestors of my ancestors had made more than a mild, as yet unrecognizable, start toward me. He has not changed much for a long time now following the winnowing away of his onetime companion genera.

He is the smallest hoofed mammal native to the Southwest, rarely weighing more than fifty pounds and usually less, standing about fifteen inches high at the shoulders, two and a half to three feet long with a third of that length contributed by his oversized snouted head. The heaviness of that head and of the short thick neck supporting it is unique, I believe, among American mammals. So too is the slenderness of his legs in proportion to the body. His twin little black hoofs seem scarcely adequate to sustain without wobbling the weight above them. Seen sideways in silhouette, he looks awkward and unbalanced. And yet, somehow, the totality of him imparts an impression of compact efficiency — an impression enhanced when he is seen in action.

His general color appears to be gray, a salt-and-pepperish gray, result of the fact that his hair bristles are banded black and yellowish white. Those forming his collar stripe simply have more of the light colored bands. A crest or mane of longer stiffer bristles runs from the crown of his head to his rump and rises erect when he is startled or angry, presumably to give him a larger, more formidable aspect. Frequently the hair on his front knees is worn away and calluses have formed there, probably caused by what I would call an occupational disadvantage of his overall structure. His short thick neck and big head give him strong leverage for rooting in the ground with his piglike snout. But the very shortness of the neck and the angle at which it sets the head on the shoulders makes it impossible for him to root very deep without dropping down on his knees. When food is scarce and he has to scrounge hard for it, he gives those knees considerable wear.

He is definitely gregarious and moves about in a group of mixed sexes and ages. What should such a group be called? The usual term in naturalists' accounts and hunters' tales is "herd." That is more accurately applied to groups of true ruminants. Surely the peccary deserves something more specific, more personal. The lion has his pride, the baboon his troop, the wolf his pack, the fox his skulk, the seal his pod, the pig his drove. The peccary is as distinctive in his way as any of them. I am indebted to B. F. Beebe for the term she uses in her *American Desert Animals,* which seems to me appropriate: sounder. No matter that this was once a Middle English name for a group of pigs. It is obsolete nowadays in that connection and the peccary can lay claim to it.

The members of a peccary group are quite vocal. As they wander about undisturbed, they carry on almost continuous conversations in barks and grunts and squeals, like a mutual reassurance society. When startled they emit loud warning signals. In flight they grunt with each jump of their seeming stiff-legged gait. Moreover they do quite a lot of teeth gnashing or rattling, under normal conditions presumably to sharpen those capable canines but when confronting an enemy to display defiance and warn against attacks. No doubt about it, for not very large animals subject to predation they are an unusually noisy crew. The term is apt; they form a sounder.

In the old days before we humans began monopolizing more and more of his territory and devising more and more efficient weapons, the collared peccary was numerous in the Southwest and a sounder might tally forty to fifty or more individuals. As late as a century ago he ranged here as far north as Oklahoma and Arkansas. Nowadays he is found only in southern portions of New Mexico and its neighbors Arizona and Texas, and the average sounder tallies no more than five or six. As a matter of fact he became almost extinct in the United States a few decades ago.

He had survived all manner of predation in the past, all the carnivore experiments evolution pitted against him — in relatively recent times, for example, that of the jaguar, deadliest of the big cats, who used to range up into the Southwest too, apparently to dine on him. But he could not survive here unaided against increasing human predation.

We egocentric humans might say that was his own fault. He had become gastronomically and economically attractive, his chunky carcass packed with good meat, his hide of fine quality, thin and strong and well marked with bristle pits, even better than the original for "pigskin" jackets and gloves. Moreover he offered good hunting experience for people so inclined, earning a place among larger and more formidable animals as "The Smallest of the BIG Game." Ranchers and others in his territory often used some of their spare time to go gunning for him simply for target practice. Some more adventurous developed a so-called sport that intrigued Theodore Roosevelt, who wrote in *The Wilderness Hunter:* "I should like to make a trial at killing peccaries with the spear, whether on foot or on horseback, and with or without dogs." Many a hunter went after him simply for his head to have it mounted as a wall trophy. A competent taxidermist, by wrinkling back the snout and opening the jaws to expose the canine teeth, could give such a trophy a sinister and savage look — precisely as could be done with a human head.

By the 1930s with depression spurring hide-hunting and food-hunting the peccary had almost disappeared from the American Southwest. Then in 1937 New Mexico, soon followed by Texas and Arizona, accorded him official protection by declaring him a game animal. Here in New Mexico he was so far gone that the Fish and Game Department gave him a "population rebuilding period" of twenty-five years before starting to issue annually a limited number of hunting permits. Again, following the rough winter of 1967–68, which took its own toll, another rebuilding period through 1971 was maintained.

A reasonable verdict would seem to be that the protection given him by the three states has been successful to date. He is still with us and his populations have definitely increased from the 1930s' lows. There is even occasional evidence that he is reclaiming a few portions of his former range.

I note with pleasure that a new reason for such protection is being cited these days. The lesson is being learned that he is economically more attractive as a living animal than as a dead carcass. He has a strong liking for the succulent interior of various cacti, particularly the prevalent and obnoxious prickly pear. He is expert at uprooting such and making meals of them, equipped as he is with competent snout and capable biting apparatus and digestive system apparently not bothered by spines and bristles. He can do as good a job of keeping down cacti on grazing lands as any manmade device — and he demands no pay for doing it. Ranchers in various areas are beginning to have a high regard for him as a volunteer employee. Some have urged that efforts be made to restore him to much of his onetime range.

It is time and enough, say I, that some of his better qualities be recognized. Too long he has had the reputation of being a mean-tempered, ferocious, dangerous animal, based primarily on such accounts as that offered in 1831 in *The Personal Narrative of James O. Pattie of Kentucky* who had been beaver trapping through much of the Southwest. Pattie's opinion of the peccary was an interesting mixture of fact and fancy.

> In these bottoms are great numbers of wild hogs of a species entirely different from our domestic swine. They are fox-colored, with their navel on their back, towards the back part of their bodies. The hoof of their hind feet has but one dew-claw, and they yield an odor not less offensive than our polecat. Their figure and head are not unlike our swine, except that their tail resembles that of a bear. We measured one of their tusks, of a size so enormous, that I am afraid to commit my credibility, by giving the dimensions . . . They have no

fear of man, and that they are exceedingly ferocious, I can bear testimony myself. I have many times been obliged to climb trees to escape their tusks. We killed a great many, but could never bring ourselves to eat them.

When I was a boy a good half century ago tales of an enraged sounder ganging up on an unlucky hunter, pursuing him with murderous intent, forcing him to climb some scrawny tree, keeping him there for hours with the tree trembling from attempts to uproot it, were a staple of the more exotic dime novels. My original notion of the peccary rested on such reading fare — though even then I wondered why the murderous intent was always presented as one-sided and no mention was made that the hunter had involved himself in the predicament precisely because of his own desire to kill — and as in Pattie's case not even for food. Very likely I also associated the peccary with the European wild boar of other tales, the much bigger and big-tusked wild razorback. When I finally met the peccary in person, I was surprised at his relative smallness, his somewhat and somehow appealing comicality of appearance, and his seeming contentment with life — if left to live it in his own way.

Even today many people with whom I talk about him still believe firmly in his nasty disposition and ferocity. No doubt there was some basis for such belief in the past. The individual peccary, brought to bay, is a capable and courageous fighter — as any dog who has tackled him can testify. When multiplied into a sounder he can be distinctly formidable. There are authentic accounts, from the old days when sounders were large-scale, of dog packs put to flight leaving dead behind them and even of jaguars forced to go without dinners. Cowboys who thought it might be fun to rope a peccary often had second thoughts when the legs of their horses were severely slashed. Moreover, in early settlement days in the Southwest, the sounders had had little experience with human predators and had not yet learned to fear them.

The situation is different nowadays, at least in my part of the peccary's range, and has been for quite a while. As witness I cite Jack O'Connor, who stated nearly thirty years ago in his *Hunting in the Southwest:* "I have seen some thousands of javelinas in my time, in

Arizona, Sonora, and Texas, and without a single exception, all of them were convinced that *homo sapiens* was a bad *hombre* and one to be avoided."

The above is not intended to indicate that an encounter with a sounder in the wild carries no element of danger — at least of possible difficulty. The chance is there, but not because of murderous intent or innate ferocity — at least on the part of the sounder. It is the result of the peccary's poor distance eyesight (he depends more on his senses of smell and hearing) and his habit of instant decision. When startled he does not stop to determine the source of the alarm, but snaps into immediate flight in whatever direction first occurs to him. Usually that is at some angle away from the source, but sometimes not. The speed with which all the members of a sounder can disappear from view is really remarkable — a speed not so much of actual running as of instant action and dedication to flight. "Shot at," O'Connor wrote, "a javelina thinks more of his hide than he does of his integrity." But anyone who happens to be in the path of his heedless flight and fails to get out of the way fast enough may discover how efficient those canines-still-canines can be.

Like most gregarious animals the collared peccary is sexually promiscuous and there seems to be a general sharing of sexual favors within a sounder. Because of the warm climate of his range, mating may take place any season of the year and thus a sounder is apt to include youngsters of various ages. The gestation period is about four months. When a female's time has come, she goes off by herself to some safe place to give birth, almost invariably to twins. Within just a few days they are able to frisk about and she takes them with her to rejoin the group. I would say she should do so proudly — and maybe she does. A baby peccary is one of the most attractive little creatures on the face of this ancient earth. The little body that with adulthood will appear awkward and badly proportioned is neat and trim and nicely balanced. The coat that will become a mediocre grayish is a glossy auburn with the species collar standing out clear and bright. The slender legs that will seem to be too thin are just right and just the right length and end in tiny jet-black hoofs. And the whole of the little creature is infused with that special attribute of the young of most hoofed mammals — a joyful appreciation of the miracle of being alive.

At this stage a young peccary is willing to be friendly with the whole wide world — even with those domineering masters of creation, us humans. He can be easily tamed and makes a fine pet, as intelligent as any dog, as fond of play as any puppy. In some Mexican villages he can occasionally still be seen tagging a child-master about, always ready for a romp. As he matures, however, the call of the wild comes ever clearer to him, the urge of his ancestry ever stronger, and he usually becomes ever more stubborn, intractable, unpredictable. Perhaps he realizes that he is no longer attractive enough to draw affectionate attention and perhaps understands that the time approaches when he may be consigned to the family stewpot. The truth more likely is that without knowing what is the matter with himself he wants the companionship of a sounder, the solace of mates, the independence of free foraging in those superb southwestern sun-swept arroyo-broken cacti-studded distances that I firmly believe can give a measure of meaning to the life even of a New World pig.

Order: *Artiodactyla*
Family: *Antilocapridae*

An American Loner

THOUGH THE PALEONTOLOGISTS, who study fossils and ancient life forms, still have their arguments, the preponderance of the probability is that all of my evolutionary ancestors along the line of descent (or ascent) from the jawed fish who became the first land vertebrates have been flesh eaters. More accurately, most of them have been as I am omnivorous in the sense they could and would eat almost anything they were able to digest, but almost certainly the carnivorous habit held throughout. Along the early mammalian way there were others with pacifist tendencies who insisted on being strict herbivores, plant eaters. They flourished for a time, then died out, while the flesh eaters continued toward me. The obvious moral is not that the meek shall inherit the earth.

Only in more recent times, though still remote, long after the distinctive aspects of mammals were well established, did new herbivores develop leading to those known today. They worked hard on the adaptations needed for success as confirmed vegetarians. New-style teeth had to be developed, new-style digestive processes, new-style defenses against predation such as size and strength and horns and antlers and fleetness, which in turn required rather drastic changes in body and bone structure.

My favorite among them is one who is a true native American, who to me is and always should be one of the prides of our continent, the pronghorn.

Like the peccary, he is an ungulate, a hoofed mammal. Like the peccary, he is an artiodactyl ungulate, an even-toed hoofed mammal. Unlike the peccary, he has gone the full ruminant route and thus is in the artio suborder Ruminantia, the cud chewers with departmentalized stomachs. Narrowing that down, he is in the superfamily Bovoidea, grazers with high-crowned teeth, chiefly plains dwellers, almost all having true horns. Down to that point he has many relatives around the world. But from there on he is completely on his own.

In his way he gave the taxonomists almost as much trouble as did the platypus. All the first people to report on him in print (Lewis and Clark led the parade) simply called him an antelope and let it go at that. When those more scientifically minded gave him attention, they bumped into difficulties. Though as early as 1818 one of them had used the label *Antilocapra americana* (antelope-goat American), on into the 1850s and 1860s he was still being given labels spluttering with such terms as *Dicranocerus, Hamatus, Furcifer, Maxima,* etc. He simply would not fit wherever they tried to put him. Like a proper bovid, he had horns. Horns, not antlers. But he defied bovid tradition by being the only one anywhere with branched horns — and, ultimate rebuff to all his more conservative bovid relatives, he treated his horns like antlers and shed them every year. There were other items to pester the taxonomists such as a gall bladder, which other bovids would say he should not have, and no dewclaws, which other bovids would say he should have. The classifiers finally accepted his singularity, made him *Antilocapra americana* again and recognized even that was not enough and gave him his own family Antilocapridae.

There is only one species of him. He is the one and only living representative of his genus, of his entire family. And he is completely and absolutely American.

The paleontologists have given him his pedigree and patent of citizenship — and so doing have solved the problem of his aloneness. Historically he was not such a loner. There have been many antilocaprids, all of them loyal Americans. The earliest developed here in

North America in the Miocene, some twenty million years ago. On through the Pliocene they were busy creating variations. On into the Pleistocene, say two million years ago, they were doing very well, many species and all trying experiments with horns, some four-horned, some even six-horned, some with spirally twisted horns — reminiscent of what the African antelopes have been doing in more recent times. Among them the pronghorn had already established himself and his personal peculiarities.

Apparently neither he nor any of his cousins from the very beginning of the line, unlike such restless hoofed mammals as the horse and the camel who also originated here, ever migrated to another continent, not even when the Bering Strait crossing into Asia and the Central American into South America were available. They liked it here. They stayed — all of them.

They had survived the extinctions of mammalian species that occurred in the Miocene and the Pliocene and the early Pleistocene. Then came the great, still mysterious die-up of the late Pleistocene, not much more than ten or eleven thousand years ago. Some 70 percent of all species of large (one hundred pounds or more) North American mammals died out. All of the major ungulates with the exception of the muskox (an immigrant from Eurasia) disappeared from the continent. All the others, that is, except the pronghorn. He came through.

He is the one remaining member of his once numerous family. He is the one and only surviving native American bovid. He was here when the muskox arrived. He was here when the modern bison came as immigrants. He was here when mammoths and mastodons and giant sloths and dire wolves and saber-toothed tigers roamed our continent. He was here when the Indians crossed into Alaska and drifted southward and he entered into their folklore and gave his name to clans in tribe after tribe. He was here when the Spanish pushed northward from their holdings in lower Mexico and gave him the name of *berrendo*. He was here supplying the tastiest of American meats to us Anglos as we conquered our way from coast to coast. Far more than any of us posturing featherless bipeds who nowadays give him "protection" so that there will be more of him to kill, he is a true American.

Though some of his early cousins roamed eastward, several of them all the way to Florida, he was and has remained primarily a western American. His modern range before we began taking it away from him (more accurately, taking him from it) blanketed the western half of the United States and up some into Canada and down more into Mexico.

In the old days his numbers were beyond counting up into the millions, equaling and probably surpassing those of the bison. Sober estimates run to forty and fifty million. All through the prairies and plains and into the foothills and valleys of the mountains the landscape in Washington Irving's phrase was "haunted by antelopes." Then we humans came along in our increasing numbers with our increasingly efficient weapons. The pronghorn held out longer than the bison because he was much smaller with a smaller market value. But by the early 1900s perhaps 20,000 were left in the whole of the American West.

My New Mexico was doing its share in the wholesale reduction. By 1915 there were perhaps 1200 left in all of the state. His range here, once covering the entire expanse, had been shriveled down to small far-scattered pockets.

New Mexico has since been doing its share too in the preservation and protection movement. The latest presumably accurate figures I have cite about 360,000 for the west as a whole, 15,000 for this land of enchantment.

I am stretching a point, a propagandist, when I assert that he is "protected" so that there will be continuing crops of him to be reaped by hunters. That is the major reason. But most hunters, those respectful of the rules, mix in a few more motives than merely that of keeping some targets on hand. Many of them share with some of the rest of us a feeling of responsibility, of appreciation, even of affection, for our peerless pronghorn. Even ranchers are learning to be tolerant of him. He is not the competition for their cattle they used to think he was. Careful studies have shown that he is not much of a grass eater, that he prefers to graze on plants usually dismissed as weeds and to browse on small bushes. He can even eat with relish plants poisonous to cattle and sheep. He might almost be called an asset on rangeland shared nowadays with cattle as he used to share it with bison. Since his diet is so different, he can actually prosper on rangeland so overgrazed that cattle would starve on it.

He is a fascinating fellow mammal, a personality in his own right, an inalienable part of our natural environment that would be a poorer environment without him. More so than that lumbering immigrant bison and the far-traveled-and-returned mustang, he has been and remains the perfect symbol of the great open spaces we prattle about as such a distinctive feature of the American West.

His adaptation to those spaces was superb and much of it uniquely his own. In a land of often drastic ups and downs and erratic temperatures he can be quite comfortable anywhere from sea level up to at least nine thousand feet. The reason he does not bother to go any higher is probably that the vegetative cafeteria on up is not to his liking. A big help against the weather is his hair, which is pithy and holds many air cells and is controlled by a special muscle system. In hot weather he can make the individual hairs stand out, permitting air to circulate close to his skin and cool it. In cold weather he can make them lie flat, forming effective insulation. The reason you rarely if ever see a pronghorn rug or couch throw is that those hairs are rather brittle and not firmly attached. They shed out easily and he is always shedding some of them and growing more, say a complete new coat about twice a year.

In a land of great distances good eyesight is an asset. The pronghorn has the best of any mammal, quite possibly of any living thing, with the exception of some of the predator and carrion birds. His eyes are large, larger than those of much bigger animals, about the same size as those of the elephant — whose own seem small because the rest of him is so big. The pronghorn's are set in protruding sockets giving wide-angle vision. Moreover, they are amazingly powerful — the equivalent of a man with 20-20 vision using 8-power binoculars. Take a pair of strong glasses and go out into pronghorn country. Whenever you spot a pronghorn, nine times out of ten he will be looking at you — and the tenth time he may well have already seen you, checked you out, and dismissed you as lacking in interest.

A corollary to that eyesight is his signal system. Like most other bovids he has a full set of scent glands (his own variations on that too of course) and can emit a variety of messages with them. But his specialty is the flashing of his well-known white rump patches: A form of heliograph he was using for communication hundreds of thousands of years before we humans ever thought of it — and we had to make gadgets for it while he carries his equipment as part of his natural endowment, neatly doubling in purpose as part of that hair-insulation system I mentioned above.   .

In a land of great open spaces the ability to move about is not only an asset but a pleasure too, an exhilaration. The pronghorn has responded by making himself the fastest living creature on four legs anywhere in the world with the possible exception of the cheetah. And a healthy cheetah might keep up with him for some hundreds of yards and then would be done while he would just be getting warmed up, ready to keep going mile after mile. I have read accounts of people cruising in a car at sixty miles per hour and having a pronghorn not only keep pace but move on ahead and cross over. A week-old pronghorn fawn can out speed any man. A three-month can leave a good horse behind. An adult could win the Kentucky Derby before the equine entrants had much more than passed the halfway mark.

He is the perfect running machine. He started with the basic artiodactyl model that was developed very early featuring efficient limb structure designed for speed, then added his own improvements. He is equipped for maximum efficiency. Large strong lungs. Double-size trachea and double-size heart. Leg bones proportioned and set just right. Double-hoofed, not single like the horse, giving better and more flexible foothold so that he can race over rocky ground that would quickly lame a horse. No bounding, no leaping, no bobbing. No energy or forward motion lost in lifting his weight up and down. Almost level in action. Meriwether Lewis, seeing him in action, thought his course "rather the rappid flight of birds than the motion of quadrupeds." Josiah Gregg put it even better in *Commerce of the Prairies:* "skimming over the ground as though upon skates."

Standing still he is not exactly beautiful. Body rather short and thick and seeming out of proportion to the trim legs. His head, straight on, downright ugly: horns directly over the eyes, ears where somehow they ought not to be, expression rather pugnacious. As in so many other things, in appearance too he is completely himself. He does not really resemble anything else — except another pronghorn. Yes. Standing still almost bizarre, outlandish. But when he starts to move . . .

William T. Hornaday, who saw more and more varied wildlife in his time than most of us now will ever see, said it for me: "The proudest, swiftest, most graceful animals I had ever seen."

I said his adaptation to the open spaces *was* superb. Past tense. It is not now. Oh, it still is — to the spaces themselves. But they have changed. They are nowadays infested with things like me.

Through millions of years he had worked out his means of survival, his defenses against his enemies. One of them he shared with the other ruminants, the ability to grab food quickly, swallow it, then retire to a safer place and regurgitate it and give it the chewing required. But his major defense was his own — at least carried to his own excellence. Eyesight plus speed. He could spot danger far off and he could outrun anything that tried to catch him. A result was what almost everyone who has written about him has called his "curiosity."

Most animals when startled, when sensing danger, seek to hide. Not the pronghorn. He wants space, openness around him, scope for his sight and his speed. Where a deer, for example, will seek cover, will slip into any nearby brush, the pronghorn will deliberately go further out into the open. Confident of his ability to outrun any predator, he wants to see what it is that might be a threat, where it is, and whether running is in order and if so what direction to take.

That technique was his defense and worked well through whole long ages. He could outrun danger. But suddenly, with no time granted him to develop any new adaptation, he was confronted with something he could not outrun. The bullet. His "curiosity" became a deadly handicap. Half a century ago he would have become extinct if we perpetrators of the bullet had not given him "protection."

Perhaps a warning lurks there for us humans. Curiosity can be a dangerous trait. His curiosity turned back upon him and could have wrought his ruin. Our curiosity in the form of what we call science has led us to the point where it is turning back upon us, can wreak our ruin in atomic blastings or in the slower strangulation of excessive population and dwindling resources. We are protecting him against the results of his curiosity. Who will protect us against the results of our own curiosity?

Order: *Carnivora*
Family: *Felidae*

The Formidable Felids

AT THIS MOMENT this room in which I sit confronting my typewriter
is a den of carnivores. I cannot include myself in the census.
Though the aroma of beef-and-chili stew drifting in from the nearby
kitchen indicates that I too am a meat eater, the taxonomists say I
am a primate. So here I sit, a minority of one primate, sharing
quarters with a pack and a pride of carnivores: three dogs and two
cats.

I suspect they regard me as foolish. It is midday, time for *siesta*.
The New Mexican sun has melted the thin sifting of snow that fell
last night and is bestowing its blessing for our common benefit
through windows into this comfortable room. And here am I busy
doing something, expending energy under no spur of necessity on
what to them makes no sense and which at times I have difficulty
justifying to myself. Much more sensible to be doing what they are
doing.

The three dogs, one male and two females, are sprawled on a
couch, presumably asleep: their eyes are closed and the snorer
among them is gently tuning up. They are conserving energy for
truly important doings. Though thoroughly relaxed right now, they
can leap into swift and vociferous action to give warning and dash
off to investigate any strange sound which obviously they can sort

out even in sleep from the normal noises of life hereabouts. Perhaps they regard that simply as a duty. There is no doubt they consider as a positive pleasure the afternoon physical and vocal routine of greeting or warning, as circumstances dictate, every human walker or jogger and every dog and horse who comes along the bridle path skirting one side of our property or the road passing the front. Since they have many hundreds of feet to patrol, they daily rack up miles of dedicated racing along the lines.

Even more sensible, perhaps, is what the two cats are doing. They are relaxed in even more complete abandon, also asleep or seeming so. They too can sort out sounds but have no urge to be dutiful. Hearing strange ones, they merely perk up and decide whether measures for immediate personal safety are required.

One of them is a young tom, a stray recently acquired from the local humane society. He is lying on his back on my desk only a few feet from the typewriter contorted into a twist like a piece of corkscrew, a position I could never attain but which with him means absolute contentment. The clatter of typewriter keys disturbs him not at all. He is conserving energy for his afternoon and evening gymnastics when he will pop out and in through the small swinging cat-door a dozen or more times an hour, cavort about the house and grounds outside in sudden galloping spurts, climb everything climbable, occasionally join the dogs in their dashes along the fence-lines. When he wants to, he can move about as softly and silently as a feather floating; also, when he wants to, he can slap his feet down so that he sounds like a small horse running on turf and can leap from a high perch to land with a resounding thump. I am sure he does the latter with mischievous intent, preferring to do it late at night when the rest of us are asleep. Many a time I have waked suddenly, trying to focus on some sound I must have heard. I peer at the nearest of the dogs, who rotate sleeping places around the bedroom. An ear may be twitching, that is all. The sound has been heard and decoded. Nothing to worry about. Then I hear what could be a small hoofed mammal thudding along the near hall and in my slow clumsy human way I can do my own decoding — and go back to sleep.

The other cat is an old lady, so old (in cat terms) it is hard to remember she is the daughter of a former member of the household who was quite capable of defending herself against any strange dog but was at last done in by two who ganged up on her out in the open where no tree was available. Right now she is curled into a furred ball on a chair. The only sign of life I can discern (we worry about her at her age) is the slight stirring of fur-hairs where her nose is tucked against her rump. She is conserving energy for what nowadays is her major activity, invading the kitchen when any human is there and putting on vocal pressure for food. Usually she does not really want it and may not touch it. She is asserting herself in her way, asking for the one constant of communication with her humans she understands best. What she wants — and gets — is reassurance that she is still a respected member of the family.

All five carnivores inhabiting this house are distinct individuals, in part because they *are* carnivores, who often show considerable variation of personality, and also because some of the potentials of their genetic endowments that would not necessarily be evoked in the wild have been developed by existence among people. All of them in varying degrees are always interested in everything that happens hereabouts. All of them again in varying degrees have learned to understand much of what is said to them, interpreting from voice tones and recognition of certain words, usually replying in their own physical and vocal languages. All of them yet again in varying degrees are attuned to the moods of the household set by us human occupants, happy and carefree on the good days, subdued and worried on the bad ones. The major distinction here is that on the bad days the dogs are more underfoot, follow about more, seem to try to understand and sympathize, while the cats simply retire to quiet nooks for what appears to be more energy conservation.

I am tempted to say that my five carnivorous roommates symbolize the whole of the order Carnivora, since many taxonomists divide it into two superfamilies: Canoidea, the dog kind; Feloidea, the cat kind. Apparently there is ample scientific justification for making a division, but the dog-cat labeling is not very happy. Taken literally it would imply, for example, that the weasel is doglike and the hyena is catlike. Moreover, of the four canoid families the Canidae, the true doglike, is not the most important in number of genera and species and individuals. Just so, neither is the Felidae, the true catlike, among the three feloid families.

I cannot even claim that my five carnivores are good representatives of their respective families. We humans have tinkered with their ancestry much too long.

Two of the dogs are Westies, West Highlanders, white Scottish terriers, far removed from the wild dog stock from which through long millenia they have come. Back along the line their ancestors were selectively bred to be hunters adapted to the pursuit and killing of burrowing animals, particularly rats and other rodents. Among the points sought were small shortlegged size, coarse thickness of protective fur, and a reckless boring-in fighting spirit. Then, rather recently, well after the Scottish terrier strain was established, their more immediate ancestors were bred for color, for the solid whiteness that now distinguishes them as West Highlanders. Our two definitely have the characteristics I have mentioned. They are wonderfully companionable and affectionate toward us, their humans, but I should hate to be a rat, even if I matched them in size, with either one hard on my tail. They have little resemblance except in jaws and teeth to any of the wild dogs that still survive in this age of man. As a matter of fact, the male looks more like a small bear, the female more like a small shortlegged fox.

The third dog, another female, was a stray abandoned (with pups too) by her former owners before she became a member of this family. By appearance and behavior she is at least half French poodle with the remainder some form of bar sinister. Back along the poodle line her ancestors were selectively bred for hunting too — but more for the open-field ground-covering kind. Among points sought were ranginess, long legs, good jaws, and thick but tightly curled fur for protection against briers and thickets and such. Then, when French poodles began to be fashionable as house pets and status symbols, they were deliberately bred for smaller size and fancy colors. The process has reached the stage at which some are miniature nervously afflicted pathetic parodies of dogdom. Ours is about medium-sized with a good light tan poodle coat, longer bodied than the Westies, cheated some (probably by that bar sinister tinge) on leg length. She resembles even less than the Westies any of her close wild relatives. Except for length of jaw, which her curly fur helps hide, and the excellent teeth she shows when she yawns, she really looks more like a small shortlegged lamb.

All three are still definitely dogs in heart and mind — with a long generational overlay of association with humans. In some aspects of behavior and certainly in appearance, they show as much the results of man-selection as of the natural selection that produced their long-ago progenitors.

The same is not as apparent in our cats — and probably not quite as true. The young tom is plainly the more civilized of the two. Though he came to us with parentage unknown, he is obviously a Persian from a long line of pampered house pets whose ancestors were bred for fine long silky fur and quite probably for amiable disposition. His luxurious coat makes him — if frequently groomed — a beautiful animal, but it would be disdained by any wild felid. It is actually a handicap. All too easily it becomes clogged with burrs and dirt and other debris and matted into tight bunches that are painful to him and hamper movement and that by his own efforts he can not untangle or remove. He very likely could not survive on his own in any natural environment and seems to understand his dependence on human help, virtually asks for it when his fur problem is acute. Unlike Kipling's Cat Who Walked By His Wild Lone, he is infinitely friendly (without being in the least servile) and confidently expects friendship in return. He will walk right over this typewriter while I am using it, stopping midway to give a greeting and suggest we touch noses, plainly enjoys being picked up and petted, and no matter how roughly handled has never bared claws against me or my wife. Very little about him except sinuous grace of movement seems to resemble the wild felid.

The other cat, the old lady who is not always ladylike in temperament and behavior, shows much less effect of domestication and man-selection. She is no particular breed, the presumable result of haphazard mating of semidomesticated then domesticated ancestors who were tolerated (sometimes encouraged) to stay around as mousers and ratters and who inhabited barns and other outbuildings until, in relatively recent times, they were admitted into households. Her wild ancestry is probably only hundreds of years back, not thousands like that of the Persian. She is more furtive, more easily startled by sudden movements or sounds, less dependent upon human companionship, seems to have more resources within herself. When younger she would have been more capable of survival on her own. She likes attention — but on her terms and is willing to use her claws to enforce them. Now in her old age she has mellowed some, is more companionable, but the central core of personality remains. I may not have quite as much open affection for her as for the other four carnivores, but I have true admiration for her. In a genetic sense she is the purest of the five, the closest to original ancestry. To me she is the essence of cat.

She has a fine serviceable shorthair grey coat. No hiding of basic contour in deceiving shapings of fur. Watching her as she moves about, I see hints of every member of the whole family of felids. She reminds me that one of the remarkable things about them is that they have kept themselves a much more closely related group than have the canids — and as a corollary that domesticated cats have been more resistant to being molded into varying types by controlled breeding than have domesticated dogs. I believe this aging grey feline is the reason that in my current considering and contemplating of American carnivores, I have found myself concentrating on the felids.

In structure the carnivores are the least specialized of any mammalian group. As hunters and killers of many kinds of prey they have had to keep themselves capable of extremely varied movement, not adapt themselves to certain types of movement at the expense of others. At the opposite extreme are, say, the bats, who had to make drastic changes in themselves to achieve flight — or ourselves, who had to undergo considerable remodeling to achieve upright bipedal locomotion. The carnivore skeleton is still generalized, can be considered the typical basic mammalian framework.

A flesh eater has to be a bit smarter (as well as better armed) than what he eats or he will soon not be eating. All later flesh eaters probably evolved from a onetime inconspicuous group of early mammals known to the paleontologists as the miacids, who along in the Eocene fifty-some million years ago began developing better brains along with such typical carnivore teeth as stout canines for biting and gripping and sharp carnassials for shearing. By end of the Eocene thirty-some million years ago two general lines were emerging, roughly reflected nowadays in the two suborders, the doglike and the catlike, those who chase and kill and those who stalk and kill. Perhaps the most obvious distinction in the current models, easily seen in their respective tracks without any necessity of trying to examine the creatures themselves, is that the doglike have fixed extended claws whose marks show plainly while the catlike have movable claws usually leaving no imprint.

Those cat claws are almost invariably described as "retractile." More accurately they are "protractile." Normally, when their owner is at rest or moving about undisturbed, they are in the withdrawn position. A muscular effort is required to extend them, to flex them into position for use. This is virtually automatic whenever their owner is startled or intends to use them, but is still a definite muscular action. To call such claws retractile is the equivalent of saying that a mouth is closable.

According to the fossil record the felids developed more rapidly than did the other carnivore families. By time the Eocene had slipped into the Oligocene, when the ancestors of the other families were just beginning to be recognizable as such, what were definitely felids were already well established. They probably were then what they certainly are now, the most carnivorous of the carnivores, the most dedicated to the flesh diet. The basic pattern that would persist relatively unchanged down to the present was already set.

Plainly it was a very good pattern. On through the following epochs, the Miocene and the Pliocene and into the Pleistocene, millions upon millions of years, the felids continued to do very well, developing and discarding genera as conditions and varieties of prey changed but always holding close to that original pattern. They could, however, be divided into two general groups differing in life styles. One was the sabertooths, who were stockily built and relatively slow moving and who developed long daggerlike canine teeth in their upper jaws for piercing the thick hide of the large heavy prey they preferred. The other group averaged slenderer in build and perhaps with an eye to the long future were content with traditional carnivore dental armament and the kind of prey for which it was quite adequate and placed greater reliance on agility and fast movement.

If capable of such a notion, the sabertooths must have been proud creatures, undisputed masters of their time, able to bring down prey, mastodons for example, dozens of times their own size and weight. In killing such they could rarely if ever use the frequent carnivore technique of a direct bite into a vital part. But their sabers could make great slashes through thick hide, inflicting wounds that would bleed profusely, with death just a matter of time — and not too long a time. Superb weapons, those sabers, accompanied by specially hinged lower jaws to give them wide play and strong neck and head muscles to give them power. Plainly possessions to prompt pride. They served their owners well through millions of years. But pride (and biological specialization) usually goes before a fall.

As the Pleistocene came to a close, only yesterday in evolutionary terms, with us humans becoming accomplished hunters and helping the process along, the big, heavy, relatively slow-moving prey of the sabertooths became extinct — and the sabertooths did too. I am sure it is no coincidence that when the mastodons and their kind died out in Europe, the sabertooths faded away there. Just as, more recently, only some ten or twelve thousand years ago, when mastodons and mammoths and the big ground sloths became extinct here in North America, so too did the sabertooths. They were poorly equipped to capture the smaller swifter prey that remained, could not compete successfully with the other group of less specialized, more agile and cunning felids whose descendants are still with us today.

All contemporary felids, all thirty-six species around the world from the smallest domestic cat to the biggest lion, are confirmed flesh eaters, most of them completely so. The two that come closest to being exceptions are the flat-headed cat of South Asia, who has a fondness for fruits, and the fishing cat of the same general region, who dines rather regularly by preference on fish and mollusks.

Though no felid disdains lesser meals and some concentrate on such, they are all capable of taking prey as large and sometimes much larger than themselves. Since they are devoted to the stalk-and-kill, the careful approach and sudden swift attack in which the shock of impact is important, they have to be both strongly muscled and extremely supple. In general their limbs are more muscled than those of the canids to give power to the impact and their necks are heavier to take the stress of violent action of head and teeth. Their skulls are shorter and higher than those of the canids, more blunt and more domed — not to give room for more brains but to provide anchorage for stouter neck and jaw muscles. The skull structure and attached muscles are so designed that maximum power from shoulders and neck flows directly into the strain of jaw action. The determined bite of a felid is more penetrating, more potentially deadly in itself, than that of a canid of the same size. When pursuing a meal the wild felid, once an approach has been made, goes straight for the kill. The canid depends more on the chase, often a long one, wearing down the prey toward exhaustion — and if the prey is large and strong, seeking to hamstring it before closing in for the kill. Moreover the canid usually hunts in packs while with the exception of the quite social lion the felid usually hunts alone.

What I call the cat-and-mouse theme slips into almost any discussion of felid killing techniques. This is the notion that felids like to play cruel games with their victims, their small victims. Most of us have seen or at least heard of cats catching mice, delaying the actual kill, letting the victim seem to escape, almost succeed, only to be nabbed anew by swift claw-pounces. The tendency for us is to assume that this is a general felid characteristic. Some of the felids do sometimes perpetrate such seeming sport — but when this is done under natural conditions in the wild, there are two probable explanations: (1) the felid is already reasonably well fed and, like a professional prizefighter working out with a sparring partner, is not playing a game, but is sharpening his skills in the craft by which he earns his living; (2) the felid is a female giving her growing kittens lessons in what will be their adult means of livelihood. I suspect that the domestic cat with meals supplied by obedient humans and bored by such a tame existence is the felid who comes the closest to making a mere game out of the cat-and-mouse business.

In the overall view the most typical felid kill has a neatness of technique somewhat resembling a surgical operation. Felid canines are flatter than the canid, more like knife blades. They are clamped down on the victim's neck, penetrating deeply to force apart the vertebrae and sever the spinal cord. Obviously that demands an ability to find the right location, adjust the grip, then drive deep the keen scalpels — and to do this with swift precision. Right. I recently read a report suggesting that since felids have very many "mechano-receptors" closely related to their teeth, they can actually "feel" with those teeth. And their jaw muscles have an exceptionally fast contraction rate. In the instant of impact when they seize small prey or leap upon larger, their protracted claws are taking hold and those superbly powered and equipped jaws are searching out the right spot for the lethal crunch.

The operation is not always successful. If it is failing and if the frantic activity of the intended victim threatens to dislodge the attacker, he will use his suppleness to shift and strike down and around for the throat. If he can get a good grip there, searching now for the windpipe and/or jugular vein, phase two will have a good chance at being successful.

I refer to the neck-bite technique as the most typical because the smaller cats who are much in the majority in the family almost invariably use it and even the larger do on smallish prey. I suspect it was one of the assets which helped the modern felids win in competition for survival with the sabertooths, who had never learned it because it would have been ineffective with their big bulky prey — and whose lopsidedly developed jaws would have been awkward at it anyway.

The big modern cats are just as expert with the neck bite as are the small, but they know it is a chancy technique when used against the really big (in modern terms) prey they favor such as the larger ungulates (the water buffalo is a good example), who have thick strong necks of their own. When attacking such after the approach, a big cat drives in fast and hard, using the force of impact to knock the intended victim off balance and his big hooked forepaws to pull the victim down, then strikes straight for the throat grip. If the victim manages to struggle to its feet, the attacker, hanging by forepaws and jaws, can swing in under its belly out of reach of flailing fore-hooves and horns and rip with his razor-studded hindpaws while his jaws are still at work. If a good hold has been obtained, the verdict is almost certain with the victim suffocated or bled to death.

With neither technique is the kill as simple and the success as likely as I may have seemed to suggest. Considering the whole process from start of a stalk to the final attack, there are undoubtedly more failures than successes — just as with many human hunters. But felids seem to be philosophical about such things (as some human hunters I know are not) and simply move on for other tries. Their average of successes, when prey is in adequate supply, is adequate for them. The decision, success or failure in each instance, is a quick one, not long drawn as often occurs with the canid chase-and-kill style of hunting. When a felid unleashes his rushing attack, the action is so fast that usually within a matter of seconds the balance has swung this way or that.

Each within his size range, the felids are formidable animals, among the carnivores the best armed. The ursids, the bears, are the only other family with forepaws that are potentially lethal weapons. The felids have this, of course, and add an almost equal deadliness of hindpaws. Because they are tuned to exceptionally swift reflexes and because their hunting style demands explosive action and instant decisions, they are inclined to be easily triggered into use of their weapons. And this, I believe, has much to do with their mating habits.

As predators who are not often prey themselves, they do not need to be constantly vigilant, can take their time with the mating process. Lack of necessity for hurry may explain why ovulation, shedding of eggs from the ovaries, does not take place in the female until she is stimulated by copulation itself. What the felids have had to develop are ways of blocking, during the rising emotions and tensions of courtship and copulation, any tendency for mating to turn into fighting.

From the male's point of view the female could be considered perverse in that she becomes attractive to him well before she is ready to mate. That is probably her method of selecting a proper consort, usually attracting two or more males who engage in mutual bluffing or actual battles until one has established superiority. Then begins the courtship, actively started by whichever of the now contracted connubial couple is the more eager and resembling the fight-play of kittens — with the significant difference that it is being staged by two fully adult highly efficient killers. At least two safeguards have been evolved. One is the common animal-kingdom inhibition ingrained in the male against any real attack on a female. The other is a ritualization of the fight-play, which prescribes that seeming attacks shall be made only on portions of either's anatomy where serious damage is not apt to be inflicted.

In the usual courtship, unless selection battles have been numerous and prolonged, the male initiates the advances — and is repulsed with what appears to be vigorous indignation. As time passes the female's behavior gradually changes. She now makes advances, seems to be inviting him — but when he responds, she still repulses him. All this while it is as if they are taking turns in arousing each other to a high pitch of emotional tension. At last she is ready — well, almost ready — and with masculine insistence he mounts her, grasping the fur of her neck in his jaws, treading down by her hindquarters with his hindfeet. That turns the trick. She assumes the proper posture under him, flat to the ground with rump slightly raised and tail swung aside. And now, on through the act of copulation, which may last a while, he continues that curious treading with his hindfeet.

The mounting is unlike the usual mammalian. He does not grasp her with his forelegs: that would be too dangerously similar to the procedure of a hunting attack. Instead he is supporting much of his weight with his forelegs resting on the ground on each side of her. The treading he is doing with his hindfeet keeps him off the balance he would normally have for an attack. Whether he knows it or not, he is deliberately keeping copulation, which all too easily could resemble an assault on prey, from becoming that. Moreover, her position, almost fully prone on the ground, is unlike that of the usual prey and hampers any forward movement she might make, which could be mistaken for attempts at escape. I think it is significant that among the big cats, potentially the most lethal, male treading is most pronounced and males withhold the bite at the neck until the moment of ejaculation.

It has happened. The two of them have negotiated an imperative yet dangerous ritual that is not invariably successful. There are rare recorded instances in which something has gone wrong and the mating has turned into a killing. But this time all has gone right — and she is the one who celebrates the more vehemently. She emits a positive scream of high excitement, pulls free from beneath him, perhaps strikes at him but plainly not in anger, rolls on the ground, licks herself thoroughly, rubs her head against anything available — and before long is making advances again. He obliges. But as time passes and she continues to seek repeat performances, his eagerness of response begins to wane. Eventually he may slip quietly away, avoiding her. Sometimes before her estrus is over she may copulate with another male. I doubt that will alter the genetic consequences of the selection preceding the original pairing. Her first consort this season will most probably have sired the kittens she will now produce in her proper time.

For most felids those kittens will be born furred but with eyes not yet open. Their mother will take good care of them and continue to do so for a relatively long period after weaning because she will have to do the killing, the providing of meat, for her youngsters while they are growing and acquiring the hunting skills of their kind, many of which have to be learned. For this reason wild felids are not nearly as prolific as domestic cats, most producing only one litter a year and the big cats only one every several years.

Though felids have strong biting power, they do not have much chewing power. Meat does not need much chewing and as the most consistent of meat eaters they have concentrated on those teeth that can be best adapted to such a diet, the canines and carnassials, have discarded some of what are the usual chewing teeth, premolars and molars, and have paid little attention to keeping good grinding surfaces on the remaining. They are even more addicted to gulping chunks of meat than are the canids.

As meat eaters they are well up the food chains of their habitats, dine so to speak on food that has already been at least once digested to be made into meat, and so they do not need either complex or lengthy digestive tracts — as do, for example, the herbivores. In general the felids hold the lead in this too. For the canids the full length of the gut is quite short in mammalian terms, only about five times body length — but for the felids it is less than four. Since meat is rich in protein and assimilation of protein produces much urea, they could have the nuisance of having to urinate often. They have solved that by becoming able to concentrate their urine to quite a degree, about five times more than can we humans.

Normally felids urinate in a squatting position, but they are able to expel urine backward when they want to use it for marking, equivalent of dogs raising legs alongside bushes or trees. Anyone who has shared a house with a healthy unaltered tomcat knows how effective for scent signaling that practice can be. Felids use it for mating-time signals and to some extent for territorial markings. It was this ability to urinate backward which led Aristotle (using logic instead of observation) to assert the silly notion that cats have to copulate backward — face in opposite directions and bring their rumps together.

Except when competing for female favors felid males are territorially fairly tolerant, individual territories often overlapping with various pathways used in common — but not at the same time. The signaling system enables them to avoid each other by what seems to be mutual consent. Females, however, are usually intolerant of each other, especially when caring for young, with the size of the territories they try to control dependent on the prey supply and the number of young to be fed.

Like most fur bearers felids have no sweat glands and yet, despite what would seem to us such a handicap, many of them live in tropical regions. Actually they use the same cooling system we do, evaporation of moisture, but they do it as do all carnivores by panting, by evaporation from the tongue, the mouth, the nasal passages. Watching my five carnivores on a hot day I used to think that the energy expended in such rapid panting must offset the cooling achieved. I know now they are not expending much energy, have their technique for holding it to a minimum. They pant rapidly because that is the easiest way. They time the panting rate to the resonant frequency of their respiratory systems, which is about 200 per minute for canids, about 250 for felids. Air flows in and out almost as if of its own volition. And they keep the flow smooth and regular by breathing in through their nostrils, out through their mouths.

All felids, in the old phrase, can "see in the dark." This is not exactly accurate. They could not see in true darkness but they definitely can see in what is the dark to us humans. They have eyes proportionately the largest of all carnivores and these are equipped with what is called the *tapetum lucidum*, a special layer at the rear of the retina that reflects light back through it for a second chance at being absorbed. This is what causes the well-known eye-shine so frequently reported by campers aware of some felid prowling about their campsite. It is also what helps felids to see quite well at a light level only about one sixth of what we humans need.

Tests have shown that they are capable of some color discrimination. They have impressively acute senses of hearing and smell, able to detect sounds through a much wider range and particularly sounds of higher frequencies than we can catch and to detect and identify scents infinitely too slight to affect our coarse blundering noses. The sense of touch in their wonderfully sensitive whiskers more than matches that we carry in our fingertips. But we surpass them in the matter of a sense of taste, probably because of our more varied diet. Tests have shown, for example, that they have no taste for sugar. They do not dislike it; they cannot taste it.

Being solitary hunters, they are far less vocal than the canids, silent most of the time except during the mating season and when a mother and young need to communicate. The lion, most social of them, is naturally the noisiest. Sometimes others of the big ones resort to roaring when out hunting, possibly to startle potential prey into betraying its whereabouts — and perhaps again just for the hell of it, releasing some energy in expression of the pride and power in them.

What has long puzzled me is the habit most of them have, almost invariably when making a stalk, of undulating their tails. I have come across various theories on that, but none satisfy. The closest to the probable, I think, is that the movement is an involuntary means of easing a bit the tension of the stalk, which demands the clamping of strict control over eager body muscles until the instant of explosive action.

Though the modern felids are the most varied and complex in coloration of all the carnivore families, under the skin they are the most similar to each other. Their differences are primarily of size and habitat. Any two of approximately the same size, though they live and their ancestors have lived thousands of miles apart, will be amazingly alike. Present even an expert with a lion's skull and a tiger's and he will have difficulty deciding which is which. The basic model worked out in the Eocene was obviously designed for long term suvival under vast shiftings of conditions and has required, I believe, fewer important later adaptations than have the beginning models of all other current medium-to-large mammals.

The very fact that all contemporary felids are so similar has given the classifiers material for disagreement, for debate over fine points at the genus level. All whose lists I have consulted agree on one item: that the cheetah, who along with other peculiarities breaks the family rule and has permanently extended claws like the canids, deserves his own genus, *Acinonyx*. Then comes disagreement, as usual between lumpers and splitters. Some taxonomists put all the rest of the world's felids into one genus, *Felis*. Some take size as a criterion and keep the smaller felids in the genus *Felis* and put the larger in another, *Panthera*. Some pick out the lynx and the bobcat (both have mere nubbins of tails in felid terms) and give them the genus *Lynx*. Some do even more splitting, citing separate genera for several of the rarer Old World felids.

The *Felis-Panthera* division seems to me to make practical sense and has a long tradition behind it. The two labels were Latin words which the Romans used to differentiate between small and large cat-style animals. There is also another difference which can underline the distinction. The small felids, when they have spots, have solid spots; the large, when they have spots, have rosettes. It seems to me, too, that the lynx and the bobcat with their distinctive tail tailoring deserve their own genus — and the label used, *Lynx,* has its own tradition, was the Latin name for the lynx, who the Romans recognized was not exactly a *Felis* or a *Panthera*.

And so, admitting that anyone who wants to file a brief can argue that all of them should be lumped into *Felis,* I split the 12 species of New World felids into three genera: *Panthera* and *Lynx* and *Felis.*

Another kind of splitting can be found in the medieval bestiaries, a splitting of one animal into two, that one the big cat we know nowadays as the leopard. Being a felid and thus a formidable fighter, he had a fearsome reputation. But Aristotle had written and such eminents as Pliny and Plutarch had echoed that it also had a sweet odor that it used to lure prey. Since the Romans (before them the Greeks) had two names for such a creature, pardus as well as panthera, the medieval scriveners intent on moralizing had no trouble concluding there were really two beasts, the pard and the panther, the one fearsome and the other noble. So for them the pard became a swift and bloodthirsty beast that struck its victims dead with a single leap — and associated with it was the leo-pard, the result of adultery between a lion and a pard. The panther, however, became an exceedingly handsome and gentle fellow with just one enemy, the dragon. After a full meal (the bestiaries never mentioned of what) the panther would retire to its cave and sleep for three days. When it awoke it would give a great roar and a wondrously sweet odor would come from its mouth. Other animals, hearing the roar, would follow the sweet odor — but not the dragon, which would slink into its hole and "lie as if dead lest the odor strike it."

Transcribers of the bestiaries liked this conceit so well that they expanded it until it was the longest of their allegories with the panther a symbol of Christ, who arose after three days to draw mankind to him, and with the sinister dragon a symbol of the devil. Even the panther's spots came to signify the many qualities of Christ.

Splitting the leopard into two animals had a long vogue, lingering until quite recently in books written by big-game hunters. Occasionally a leopard has an excess of melanin in its coat, the dark brown pigment that in various concentrations gives the yellowish and brownish colorations. The result is to give such a one an overall dark brown-to-black appearance. Mowgli's friend Bagheera in Kipling's *Jungle Books* is one such. The difference is simply a color phase within the species, yet for a long time many people insisted such dark leopards constituted a separate species and were properly panthers.

Curiously enough that word "panther" has been a favorite of poets but used by them chiefly in the bestiary "pard" sense denoting sinuously sinister and bloodthirsty qualities. For my part I have had more than enough of the word and would be willing to drop it out of the language, keeping only the original "panthera" as a genus label. It has been used too often as a catchall name, slapped on any catlike animal who might be fairly large or only looked so to people catching a sight of it. At one time or another I have found every one of our North American felids described as a "panther." The word has been tossed about so indiscriminately that when in my reading I come on a reference to a "panther," I have to hunt hard for other hints as to what felid is meant.

Of all twelve New World felids only one is a *Panthera* qualified for grouping with the lion and tiger and leopard of the Old World. That one is the jaguar, name direct from the South American Indian "jaguara." In weight at least he is the king of American cats, females averaging close to two hundred pounds, males well over two hundred with occasional individuals approaching three hundred. Such figures can be misleading as averages often are. While there is only one species, *Panthera onca,* some subspecies differ substantially in size — and sometimes a large subspecies and a smaller inhabit adjoining areas apparently without interbreeding, which could mean that a separate species is in process of emerging.

He likes a warm climate and when the Central American land bridge became available was one of the early and successful pioneers out of North into South America. By time Columbus launched the European human invasion of the Americas, the jaguar ranged from what is now southwestern United States all the way down to south-central Argentina. In fact, he was much more numerous and widespread in South than in North America, no doubt because in addition to a warm climate he also likes well-watered habitats and moist tropical forests and even marshlands, and South America is more generous in providing them. By confining himself to the Southwest in this country, however, he gained a certain distinction I claim for him, that of being the one major United States game animal NOT killed by that archpredator Theodore Roosevelt. "It has been my good luck," wrote T.R. in *The Wilderness Hunter,* "to kill every kind of game properly belonging to the United States; though one beast which I never had a chance to slay, the jaguar, from the torrid south, sometimes comes just across the Rio Grande."

I suspect that T.R. played down the jaguar's range to play up his boast. Not too long ago the jaguar did more than "come just across the Rio Grande." Spanish Mission reports indicate that he used to roam as far north in California as the San Francisco Bay area. In Arizona he was found as far north as the Grand Canyon Rim, in New Mexico up the Rio Grande valley into the northern mountains, and he was well known in the brush country of south Texas. In T.R.'s time he was still definitely an inhabitant of the United States. But now no more. The only jaguars resident in the United States nowadays are in zoos.

I write that last though some jaguars were very recently killed here in New Mexico. A contemptible creature in the shape of a man brought them into the state in cages and released them for high-paying clients called trophy hunters to gun down.

The jaguar resembles the leopard more closely than he does any of the other big cats, but there are differences. He usually has fewer and larger rosettes (often with a spot or two within them) and his chest rosettes tend to elongate into dark bars that are almost stripes. His larger subspecies are a fair margin larger than those of the leopard, more powerfully built with thicker shoulders and chest and barrel and more massive head and jaws. He packs his weight into a more robust, more compact, more strongly muscled body than does any other felid — and does this without losing much if any of the family sinuous agility. To me, watching him move, though I have done that only in zoos, he imparts more of an impression of concentrated power than any other animal of comparable size.

No. I was once given an even stronger impression by another carnivore, some years ago when I sat on a grass hummock for several hours somewhere up in the Dakotas watching a grizzly bear pace back and forth inside an old freight car from which one side had been removed and replaced by bars. This was not a particularly big grizzly, probably not yet full grown, who had been captured only a day or two before. He was lean and hungry, looked as if he had not had a good meal in weeks. He paced there, twenty-seven steps and turn, twenty-seven steps and turn, endlessly, ignoring me not fifteen feet away, looking into the distances beyond me. We were silent, both of us, the only sound the scritch of his permanently protracted claws on the floor of the freight car. More and more awareness seeped into me of how life pulsed in him, how power flowed through him. Unfavorable memories wandered through my mind of other grizzlies seen in zoos, well fed and fat and bored into stupidity. I recall thinking *this* is the real thing — and *that* is what we will do to him. Never since have I had any real fondness for zoos.

That was a single instance never repeated with a grizzly, undoubtedly because I have never seen another in or fresh out of the wild. But every jaguar I have ever seen has given me the same impression. The jaguar is the very essence of concentrated physical power. He is our leopard and tiger compacted into the one beast. Aldo Leopold, writing of a camping trip in the delta of the Colorado River, summed what I am trying to say far better than I ever could.

> At every shallow ford were tracks of burro deer. We always examined these deer trails, hoping to find signs of the despot of the Delta, the great jaguar, *el tigre.*

> We saw neither hide nor hair of him, but his personality pervaded the wilderness; no living beast forgot his potential presence, for the price of unwariness was death. No deer rounded a bush, or stopped to nibble pods under a mesquite tree, without a premonitory sniff for *el tigre.* No campfire died without talk of him. No dog curled up for the night, save at his master's feet; he needed no telling that the king of cats still ruled the night; that those massive paws could fell an ox, those jaws shear off bones like a guillotine.

The jaguar differs from the leopard and tiger in one aspect of character. He is not as ready in response or as aggressive in confrontation with humans. Despite his size and deadly armament and amazing strenth he attacks (more accurately fights back) only when at bay, prefers if possible to slip away. Nonetheless the hunting of him has always involved the chance of serious danger. When he does attack, he does so with full impetus of all the speed and power in him. Since that is usually from only a short distance away, out of a thicket or clump of brush, the hunter using a rifle has scant scope for more than one shot and if that is not fatal is in for trouble. Sascha Siemel, renowned in South America as a jaguar killer, always insisted that a special spear (the biggest and most formidable spear I have ever seen pictured) was the best weapon. In the few flashing seconds of the onrush he would plant the haft end of the spear in the ground with the big pointed and razor-edged blade aimed forward for the jaguar to impale himself on it — and as long as he, Siemel, retained hold of the haft, he was still armed to finish the fight. Even so, I note in reading of his exploits that he had to have human beaters or dogs to bring the beast to bay for him — and then had to irritate and anger it into attacking. Invariably he, not the jaguar, was the real aggressor.

Other reports through several centuries now of actual unprovoked (even provoked) attacks on humans are extremely rare and difficult to verify. They are so rare that the same few have been constantly repeated and elaborated. A good example is the one Darwin cited in *The Voyage of the Beagle.*

> I was told that a few years since a very large one found its way into the church at St. Fe: two padres entering one after the other were killed, and a third, who came to see what was the matter, escaped with difficulty. The beast was destroyed by being shot from a corner of the building which was unroofed.

Darwin heard that tale in the early 1830s. In 1859 it appeared as a longer and bloodier account in a government report on "The Mammals of the Mexican Border Survey." By then the death toll of humans had risen to four. I came on it again in James Greenwood's *Wild Sports of the World* of 1870, this time not so bloody but embellished with imaginative details. The jaguar had become "very large"; the two padres killed had become "very corpulent" giving the jaguar a "strong clerical taste"; the third padre had become "of the slender order" and thus able to dodge and escape; the jaguar was duly slain for his "sacrilegeous propensities."

In *Mammals of New Mexico,* 1931, Vernon Bailey cited it — and rated it "very improbable." I have found it or references to it in various other books and twice I have had it told to me by native New Mexicans, each time with the human death toll compromised to three, and each time supposed to have taken place in a different New Mexican town.

Praise him or blame him, though physically more than capable of it, the jaguar has not provided us of the New World through several centuries with enough of a man-killing record to match just a few years of that of any one of his panthera-class Old World cousins.

The two bobtails, lynx and bobcat, were holdouts when the Central American land bridge was restored and North American cats began crossing into South America. Obviously these two preferred temperate cooler-to-cold climates. No subtropics on into tropics for them. They were too busy populating all of North America coast to coast from well down in Mexico all the way up to the northernmost tips of Canada and Alaska. They were dividing that huge range between them, the lynx taking the northern half and the bobcat the southern with a wide band of overlap between. And meanwhile the lynx, finding a crossing now and again available over what is now the Bering Strait, was extending his personal range into Eurasia, becoming the one modern wild felid who is both an Old and a New Worlder.

All that happened rather recently in evolutionary terms and the lynx of Europe and Asia is still the lynx of North America, all one species with distinctions only at the subspecies level. European lumpers, those who put all or most felids into the one genus, label him *Felis lynx.* Along with most American authorities I make him *Lynx canadensis,* which strikes me as just right because he and the bobcat deserve their own genus and *canadensis* for his species name aptly designates his species origin in what is now Canada. The bobcat, who has remained strictly North American, is usually *Felis rufa* or *rufus* to the European lumpers. I make him *Lynx rufus,* companion in genus to the lynx, and accept the species name because it fits, is simply the Latin for "red." The bobcat tends to be reddish brown in color, which helps differentiate him from the lynx who tends to be silvery gray.

The two are so much alike that they must have had a common ancestor not too far back. In the broad area along both sides of the Canadian border where their ranges overlap even an expert needs more than a quick glimpse to know one from the other. Closer inspection can reveal differences. Though a big bobcat may be as large as or even larger than a small lynx, on the average the lynx is bigger and heavier, has somewhat proportionately longer legs and bigger feet — these last serving as snowshoes helping him get about in the deep snows of his northern range. Being a northerner, he has paid more attention to his coat, has longer and thicker and softer fur and more luxuriant ear tufts and mutton-chop sideburns. His ear tufts are almost tassels while the bobcat's are more erect and pointed. He is more all over uniform in color while the bobcat usually has dark spots and bars on his legs. His tail is even shorter than the bobcat's and has an all black tip, while the bobcat has two or three dark bars and a dark tip only on the topside of his tail and these have a white edging.

Since the lynx was well established in the Old World as well as the New when evolution played the mean trick on all other creation of producing *Homo sapiens,* inevitably all manner of mixed-up semi-information and fabled nonsense would accumulate about him. For a full and fascinating account I recommend the Reverend Topsell's *The History of Four-Footed Beasts,* first published in 1607.

Topsell gave the lynx full treatment, fine-print folio pages of it. For example, he discussed at length the ancient belief that the urine of the lynx hardens into a precious stone and added that this "for brightness resembleth the Amber . . . of the male cometh the fiery and yellow Amber, and of the female cometh the white and pale Amber." Of the many other notions he cited, I pick just a few. One, that "when they (Linxes) see themselves to be taken, they do send forth tears and weep very plentifully." Another, that "of all Beasts they see most brightly . . . their eye sight pierceth through every solid body, although it be as thick as a wall." Yet another, that "all the nails of the Linx being burned with the skin, beaten into powder, and given in drink, will very much cohibite and restrain abominable Lechery in man: it will also restrain the lust in women being sprinkled upon them."

One item at least was grounded in historical fact. "The skins of

Linxes are more precious, and used in the garments of the greatest estates, both Lords, Kings and Emperors . . . The claws of this Beast, especially of the right foot, which he useth in stead of a hand, are encluded in silver, and sold for Nobles a piece, and for Amulets to be worn against the falling sickness."

I am grateful to Topsell for yet another item that is a far more accurate assessment of the true nature of the lynx than the majority of accounts through the next centuries, usually given by people who automatically assumed that anything wild and catlike was ferocious and bloodthirsty. There was an animal being kept in the beginnings of a zoo in the Tower of London in Topsell's time which was observed and described by a Dr. John Cay. From the description unmistakably a lynx. "He is," reported Cay as reported by Topsell, "angry at none but them which offer him injury . . . He is loving and gentle unto his keeper, and not cruell to any man."

I add here that Goldsmith, who condemned all catlike creatures as fierce and rapacious and cited the belief that the lynx has "of all other quadrupedes the shortest memory," did pay him one compliment. In contrast to those of the "panther kind," which to Goldsmith meant the longtailed, he has "the physiognomy more placid and gentle."

With dense fur fully four inches long in winter, with ear tufts and wide drooping sideburns, with hair mats on the sides and soles of already big feet, the lynx appears bigger than he really is. His average weight is probably no more than twenty-five to thirty pounds with forty the maximum for a very big one. Even so, being a felid and thus well armed and able when pushed to it or cornered to give a good account of himself, he has virtually no enemies but man. On occasion a puma or a wolverine might take a meal away from him — but would have a tough time trying to make him into a meal.

His range and that of the snowshoe hare coincide for the simple reason that the latter is the staple of his diet, so much so that his populations normally follow the fluctuations long noted in hare populations. Apparently what happens is that when the hare cycle is at low point, though the lynx can turn to some extent to other prey, he practices his own form of birth control by becoming temporarily unable to breed as usual. As the hare cycle swings up, his potency climbs in accordance.

That is the way the lynx and the snowshoe hare have kept themselves in an ever-readjusting balance — and would continue to do so if left to their own devices. But for quite a while now we humans have been interfering. The lynx likes forests, particularly pine and spruce forests, and we have steadily gnawed away at his habitats, pushing him further back into the remaining wilderness areas. Just as in Topsell's time, the fur of the lynx has continued to be well regarded and we have trapped him for it — and he has been relatively easy to trap. For some reason, perhaps because he is more trusting and less wary of human guile than most of his felid relatives, he seems never to have learned much about avoiding traps. Since he is a competing predator (occasionally dines on a small deer), it is only very recently with his numbers way down that we have begun even thinking of giving him the partial protection of declaring him a game animal, thus subject to controls. He is close to becoming another endangered species.

Among the taxonomic tags attached at various times to *Lynx rufus,* the bobcat, was *Felis cauda truncata,* Cat with Tail Cut Short — which he is, of course, though not so much so as the lynx, who has an even shorter tail. Every time I come on that *cauda truncata* label, I recall the Jicarilla Apache tale of why the bobcat's tail is short. That tale is fairly long and hilariously detailed and I cut it short here to the essential point.

By ingenious tricks trickster Coyote killed a batch of prairie dogs and put them in a fire to cook. Tired now, he lay down for a nap. Along came Bobcat and ate the prairie dogs. Coyote awoke, finally figured what had happened, tracked Bobcat and found the latter taking his own nap. Coyote took out Bobcat's rectum, roasted it over a fire, woke Bobcat and tricked him into starting to eat it. When not much was left, he told Bobcat what it was and Bobcat hastily put what remained back in place. In his excitement and haste he pushed in some of his tail.

Though the bobcat is a wild cat and often called so, he is not *the* wildcat, who is correctly a longtailed Old Worlder who used to be common in much of Europe and a large part of Asia and who looks like a largish domestic cat except for his tail which is heavier and ends in a broadly rounded nub instead of a pointed tip. The bobcat averages larger and like the lynx has a blockier, squarish body with longer legs and much bigger feet as well as much shorter tail. The bobcat was and likely still is an inhabitant of every one of our contiguous 48 states plus much of southern Canada and northern Mexico and is without any doubt the most numerous of our wild felids.

That last is hard to believe. He is primarily a nocturnal prowler, so elusive, so adept at keeping out of sight or out of knowledge that he is in sight, that he can be a close neighbor to a populous human neighborhood without anyone being aware he is in residence. It is easily possible for one of us to live out a lifetime in bobcat country without catching a glimpse of him. During thirteen years on a small ranch here in New Mexico with frequent excursions further into the back country where his tracks showed often, I achieved just one look at him — and from his behavior it was obvious that he had seen me long before I saw him and had dismissed me as a harmless eccentric worth some curiosity but no effort at avoidance.

Not only has he been able to adapt to living in proximity to (sometimes within the boundaries of) human settlement, but by preferring more open country than does the forest-loving lynx he has taken advantage of our human activities to extend his habitats. As we have logged away forests, leaving the kind of cut-over scattered second-growth and brushy areas he likes and the lynx shuns, the lynx has been losing ground while he has been gaining it. Throughout his range as a whole there may be as many of him right now as there were back when Henry Hudson paid tribute to him by naming those New York state hills easterners still call mountains the Kaatskills.

I would place a big bet that, if the implacable onrush of human population and attendant crowding and crushing of our fellow creatures toward extinction is not halted, the bobcat will be the last of the American felids to go.

One of his assets is a lack of favoritism in foods. Whatever is available in his territory will do as long as it is meat. He is small enough to be an expert, perhaps the prime expert, at preying on rodents down to the smallest. Having all the felid armament advantages, he is also big enough to tackle prey much bigger than himself. And that has given him a bad reputation with hunters and ranchers.

There is no doubt that now and again he manages to kill a deer and that sometimes he raids a flock of sheep. But he is far less guilty in such matters than his accusers insist, is often the victim of circumstantial evidence and human prejudice. He comes on the carcass of an animal that has died of some natural cause (or, say, a deer that has died a lingering death from a gunshot wound) and makes a meal — and is automatically convicted because his tracks and other markings are found there. I recently read a report of a study conducted in Idaho that revealed that of twenty-six deer on which he had dined twenty-three could be established as having already been dead when he came upon them. This was good bobcat country and flocks of sheep were in the area, yet during the three-year study there were no bobcat attacks on them — and checks of stomach contents and feces of some three hundred bobcats showed only one instance of sheep remains. The occasional bobcat who does take to killing sheep (or other domestic stock such as poultry) and may even, as has happened, run amok and apparently kill for the sake of killing, is, like his equivalent among humans, a fact but also a rarity. The usual bobcat more than pays for an occasional meal at human expense by his efficiency at rodent control.

In the dime novels I read as a boy when this century was young the hero was often a tough hombre who could "whip his weight in wildcats." High praise. And quite beyond possibility. The toughest hombre, using only his personal physical endowments as would be his adversaries, would have a tough time whipping one bobcat without taking on the five or six more required to match his weight. I recall what happened to the toughest hombre of my acquaintance, one who would qualify in structure and temperament and experiences as one of those oldtime heroes, when he tackled a lone bobcat. At the time he was a high-wire man stringing a power line in the back country when he saw this bobcat perched atop one of the tall poles apparently enjoying the view. Typical of him was his immediate notion that he would just grab that little ol' wildcat and pop it into a burlap bag. He felt well armored because he was wearing his usual ten-gallon hat, a lined denim jacket, stout pants, high boots, and leather gloves. He strapped on his leg irons, slapped his holding belt around the pole, and started up. The bobcat watched him come, waited long enough to make sure he intended to keep coming, then started down. Not many minutes later this tough hombre appeared at the nearest ranchhouse, ours, to clean up some and absorb quantities of his favorite beverage, hot black coffee. Where was the bobcat? "Out there somewheres, likely still laughin'. That damned little ol' sonofa climbed down me like I was part of the pole, takin' chunks outa me on the way."

For my money the bobcat is the most effective fighter pound for pound of the whole formidable felid family. His armament, even for a felid, is impressive. His claws are proportionately longer than most of the others and his canine teeth, while not as big around, are as long as those of the puma, who weighs ten to fifteen times as much as he does. And when he fights, he FIGHTS — pours into the fracas an explosive concentration of every last iota of energy in him. That cat who bursts into an eye-blurring frenzy of action in Disney's *1001 Dalmatians* is obviously, despite his long tail, a bobcat at heart. An ordinary everyday bobcat is usually more than a match for a dog many times his size and can inflict amazing damage on a whole pack of dogs before they manage, if they manage, to kill him. When cornered he will attack anything and fight to the final dying gasp.

Note the words "when cornered." He does not go about spoiling for a fight, has no interest in fighting for its own sake. If there is a way out, he much prefers to take it. Only as a last resort does he become what the traveled William Bartram called him, "a fierce and bold little animal." Normally he is a quiet elusive gentleman who asks only that he be allowed to live in his own quiet elusive way. Adopted young and treated right, he can be a fascinating pet.

Audubon and Bachman did not think so. "We once made an attempt at domesticating one of the young of this species, which we obtained when only two weeks old. It was a most spiteful, growling, snappish little wretch and showed no disposition to improve its habits and manners under our kind tuition." I know too many people who have had the exact opposite experience with young bobcats to accept that passage as anything but a statement of failure on the part of the two tutors. Their motives may have been good; plainly their methods were not.

A bobcat kitten, once he has learned he can trust humans, behaves very much as does a domestic cat kitten, is just as responsive and playful — and seems to know he has better armament because he does what domestic kittens often forget to do, keeps his claws carefully sheathed when being handled and even in rather rough play. As he grows older, he becomes more reserved, acquires a kind of dignity, is still companionable but prefers not to be petted. The relationship he wants with humans is one of mutual respect. When grown he is no longer exactly a proper member of the average household, chiefly because he does not realize his own size and strength. If cooped up now he is apt to become moody and unpredictable for the good reason that he is a bobcat and his genetically programmed destiny is to be a free-roaming hunter in the wild night. It is time to let him go and to do that in good bobcat country. There are those who have done so and later met him again to discover he had not forgotten a former friendship because he deliberately came into view to give a respectful greeting in the form of a familiar deep-throated purring meow.

I wrote back along the way there are twelve New World felids. Subtracting the jaguar and the two bobtails leaves nine, all of them in the *Felis* genus. Four of the nine are found only in South America and I know very little about them except that they are smallish and presumably are relatively recent species adaptations to some of the varied habitats encountered there when their ancestors emigrated from North America. Presumably again, those ancestors were representatives of the remaining five *Felis* felids who had developed here in North America and who nowadays inhabit both North and South America. All of these five are very much at home in the subtropics and tropics.

F. tigrina, the little spotted cat, likes such climates so much that on this continent he stays down in Central America and does not venture even as far north as southernmost Mexico. Though a trifle larger than the average domestic cat, he is the smallest of the North American wild felids. He justifies his "spotted" name by having multitudinous dark spots even on his belly and a series of dark rings on his tail.

F. wiedii, the margay, ventures much further north, extends his range throughout southern Mexico and on up the coastal areas, on the west almost to our California and on the east all the way up to the Rio Grande and a small way across into southern Texas. He is somewhat larger than the little spotted, about twice the size of the average domestic cat, rather slender in build with a longish tail, and is quite gaudily spotted and striped in his own way.

F. yagouaroundi, the jaguarundi, has much the same range as the margay and has not only crossed the Rio Grande into the southern tip of Texas but over to the west has also crossed the border a short way into Arizona. He is only slightly larger than the margay but quite different in appearance, looks more like a large weasel than a cat. His body is long, his tail almost as long, his head small, his legs short. In coloration he is the most uniform of the American felids; that is, he has no spots or stripes and is virtually the same color all over except for some white showing around his lips and on his throat. But he does have (as does the leopard) the color peculiarity of being dichromatic, having two phases. In one he is brownish gray to blackish, in the other rusty yellowish to reddish. For a time his reddish phase was thought to represent a separate species called the eyra, but it is now well established that there is just the one species with the two colorings often showing in the same litter. No matter which color he wears, he is a creature who literally can become "pale with anger." His undercoat is lighter than his topcoat and when he is aroused into bristling all over in the typical felid manner, the undercoat shows through.

F. pardalis, the ocelot, is the sartorial dandy of the entire felid roster. A quick glance at him can suggest that his species name meaning leopardlike is correct, that he is a small edition of a well-marked leopard. A better inspection reveals that he makes a leopard seem a drab dresser. His coat is rich and glossy, positively splendid in coloration, the background gray setting off fawn-colored patches and spots that are edged with black, the whole shading down to pure white with pure black spots on his underparts. He uses those distinctly edged patches and spots to specialize in imaginative designs and patterns that vary from individual to individual and even from side to side on the same individual. Beyond doubt he is the handsomest of the whole American felid tribe.

Like the margay and the jaguarundi he is rarely seen nowadays north of Mexico, but not long ago his range extended out of eastern Mexico to cover most of Texas and out of western Mexico up into Arizona. Since range plottings for both states touch the southern corners of New Mexico, I like to think that now and again he has wandered into my state.

Goldsmith called him "the Catamountain, which is the Ocelot of Mr. Buffon, or the Tiger Cat of most of those who exhibit it as a show." Goldsmith also described him as "one of the fiercest and, for its size, one of the most destructive animals in the world." Goldsmith was wrong. Goldsmith disliked all felids (rated even the domestic cat as a "faithless friend"), primarily because they did not render what he regarded as proper homage to humans and were "incapable of adding to human happiness." Moreover Goldsmith and Buffon both based their opinion of him on the behavior of two ocelot kittens brought to Europe from South America and raised and treated and kept under conditions scarcely calculated to give them amiable dispositions.

Actually the ocelot is the most amiable and philosophical of wild felids, of the American anyway, which is reflected in his facial expression, placid, friendly, almost complacent. Even when trapped, he retains much of his cool, does not snarl and spit and thrash about in fear or rage, instead seems to adopt an attitude of watchful waiting, of trying to figure what this is all about and what is going to happen next. He has frequently been at least partly and occasionally wholly domesticated and when with humans whom he knows and trusts is quietly friendly and always keeps his claws sheathed. I have heard that the artist Salvador Dali kept an ocelot as a studio pet and I recall that a few years ago a squad of Mexican Army aviators on a mission to Washington brought along an ocelot, not in Goldsmith's phrase "as a show" but as a traveling companion.

The ocelot can be up to four feet in length (almost half of that tail) and weigh up to forty pounds. As a prey killer and as a fighter he is quite as capable as any felid his size, justifying his Mexican name of *el tigrillo,* the Little Tiger. But apparently he is not as nervously tense, as quick-triggered, as potentially explosive, as are most felids. A group of him kept in a cage, a condition most felids simply would not tolerate, will get along quite well with little quarreling. He and his mate often hunt together and some naturalists insist that they live together year after year and that he helps her feed and train their young.

The fifth and last of our *Felis* felids is as much at home in the tropics and subtropics as the four I have just been discussing — and is equally at home in other regions, in fact has had the largest and most varied range of any American mammal. This one is without question the top cat of the Americas. Before we humans began crowding and killing him out of big chunks of his original holdings, he ranged coast to coast from well north in Canada continuously southward to the southern tip of South America. He was at home from coastal lowlands up to all but the topmost peaks of the Rockies and the Andes. He prowled every kind of habitat encompassed in that tremendous double-continental sweep. Two centuries ago Linnaeus gave him the scientific name that, after many other suggestions and many arguments, is now universally accepted: *Felis concolor,* the puma, the cougar, the mountain lion, the New World King of Cats who is concolor, similar in coloration to, the Old World King of Beasts.

Since it is impossible to run down all the names formal and vernacular the Old World lion has had through the years and his own once very extensive empire, I hazard the statement that our roughly equivalent New World lion (Goldsmith called him our "pretended lion") has had more common names than any other felid. I have a list of more than forty in English and have seen lists of at least twenty used in South America and twenty-five from various North American Indian languages. The three names I cited above are those most commonly used in this country and long ago I felt obliged to choose one of them. Mountain lion is the usual in my part of the country, but I reject it because he is not a lion, is distinctly himself — and though nowadays he is found mostly in the mountains that is simply because we humans have long been pushing him out of the lower levels. Buffon's entry, cougar (first spelled couguar) is better because it comes from a Brazilian name for him, but to achieve it Buffon had to do considerable contracting of the original and there is reason to suspect he was really writing about a different animal, likely the jaguarundi. Puma is my choice, direct from Peruvian Indian, first reported by the Spanish explorer Pizarro more than three centuries ago and long backed by staunch adherents. I note that naturalist Stanley P. Young and taxonomist Edward A. Goldman, when preparing their thorough study and classification of him and his subspecies in 1946 that is still the standard book in the field, chose the title: *The Puma.*

Columbus was probably the first European to put the puma on record. In 1502 during his fourth voyage along the coast of Central America he reported "some very large fowls (the feathers of which resemble wool), lions (leones), stags, fallow deer, and birds." As could be expected, Captain John Smith was the first to render a direct report in English: ". . . there be in this country Lions, Beares, woulues, foxes, muske catts, Hares felinge squirells and other squirrels . . ."

Long before the European invasion of the Americas, the puma was bedded in Amerindian folklore and usually, whether in North or South America, as a symbol of physical power and success in hunting. In the Jemez Mountains not fifty miles as one of the big black crows I see out my window could fly from here are two much-weathered pumas carved out of volcanic tufa. They are crouched within an enclosure made of rock slabs, their tails extended, heads facing eastward. In the old days this was a hunting shrine known all through the Southwest, visited not only by the Pueblo Indians of the general area but also by other tribes from far distances. Fortunately it is protected nowadays in that it is within the boundaries of Bandelier National Monument and since it is difficult of access has not yet suffered the usual tourist blight. I would not be surprised (and hope I am right) if now and again offerings of prayer sticks are still made at this ancient shrine by those who have the right to make them.

That same crow would have even less flying to do to reach Cochiti Pueblo. Recently I was reading Cochiti tales collected by Ruth Benedict and published by the Bureau of American Ethnology in 1931.

Right away in an early origin tale I met the puma — name presented by Benedict as mountain lion. Back in the beginning of things when the animals were created, the meat eaters had to undergo a four-day fast to determine how prey would be allotted to them. On the third day coyote (always a sly one) went into the outer room for a drink and slipped some sacred meal into the water he drank. He then persuaded wildcat to do the same. On the fifth day when judgment was rendered, coyote and wildcat were told that their hunting would involve "hardship and great labor" while the others would always be successful. Mountain lion, who had faithfully followed the fast, was declared "chief of all the animals."

I like that. To me the puma is chief among American carnivores. Not because he is the largest, which he is not — though he is the largest of the *Felis* felids. On the average he is surpassed in size, certainly in weight, by the jaguar among American felids and the bear among the other American carnivores. But he is the most adaptable to differing environments, has had by far the most extensive range, and is the most all-around accomplished predator — surpassed in that only by the alltime champion, man.

He has a long narrow but deep body ending in a long blacktipped tail, short but powerful legs, a longish neck and a superbly dentured flattish head that often seems small in relation to the rest of him. For all his lankiness he is a compact parcel of power and agility. With his wide range and array of subspecies he can vary much in color tones, from a rufous gray to a reddish tan, usually with a darker streak along the center of his back. The frequent family tradition of dark spots and stripes shows on him only when he is very young and fades as he grows older. Again, he can vary in adult size, from individuals only slightly larger than the ocelot up to others matching a smallish African lioness. In general the males are larger than the females and both follow the usual mammalian rule of being larger in temperate than in tropical regions and larger in the mountains than in the lowlands.

Some years ago I put a 200-pound puma into one of my short novels and was surprised when several reviewers asserted that was ridiculous, an affront to fact. Admittedly for dramatic effect I made that puma a big one, but I knew he was within the realm of possibility. I have just reassured myself again by flipping open the Young-Goldman volume, not bothering to consult tables, simply looking at a few of the photographs included. Up come pictures of a puma shot in 1921 that weighed 207 pounds, another killed in 1927 that weighed 217 pounds, yet another slain by Theodore Roosevelt in 1901 that weighed 227 pounds. And I recall reading in one of Ernest Thompson Seton's books that the heaviest "taken" in Arizona up to his time weighed 276 pounds — that after the intestines had been removed.

Admittedly again, those were all very big ones. Current averages in this country probably would not exceed 140 pounds for males, 110 for females. I suspect that such figures, if they could have been obtained, would have been much higher back before we humans began so seriously to trim down puma populations and habitats and food supplies.

The Young-Goldman volume is subtitled: *Mysterious American Cat.* The first adjective there is justified in that the puma is an animal about whom contradictions cluster and endless arguments endlessly continue. I am inclined to believe he is as much misunderstood as mysterious. So many conflicting opinions of him have accumulated that he might seem to be not one but a dozen different beasts. Through the settlement years in this country excessive sensational emotionalism, all too usual in regard to large predators, saturated what was thought and said and written about him. Later most supposed information came from hunters. All too often the hunting mentality has its own built-in prejudices — and regarding an animal as your own prey and looking at him over the sights of a gun is not the best way to obtain an objective opinion of a fellow creature. Even biologists interested in the puma tended to follow the traditional procedure of studying corpses, then adding to the dead details observations based on the behavior of those penned in zoos. It is only quite recently that dependable studies have been made of him alive and living as he wants to live in those dwindling portions of his onetime range we have not yet taken from him.

One of his "mysterious" aspects is his attitude toward us humans. Normally it is almost friendly — if we do not (but we usually do) repudiate him by yelling or throwing things or shooting at him. Even then his most usual response is to try to get away as rapidly as pos-sible. Except in rare instances he avoids closely settled areas where he might bump into too many of us and where we have eliminated most of his regular food supply, yet he is intensely interested in any individuals who invade his home territory — interested in them ob-viously not as possible prey or as enemies but as creatures who have a peculiar fascination for him. He may trail a hiker for hours and miles, watching every move as if making his own scientific survey of an intriguing specimen. He may prowl about a campsite as if trying to figure what the campers are doing and why. He has been known to stroll right into a camp as if to see what was going on. He has even been known to come up to people and act as if he wants com-panionship, wants to play or to communicate in some manner.

That attitude has prompted ingenious attempts at explanation. Perhaps the most ingenious is that back when humans were first using the Bering Strait crossing to enter and spread down through the Americas and were still hunters and gatherers and scavengers, some kind of primitive association developed between man and puma, perhaps based on the scavenging of each other's hunts, per-haps involving some form of cooperation in hunting roughly similar to that regarded as the beginning of the relationship between man and wild dog — or to that between man and cheetah in Africa. The sense of association then tentatively developed, so runs the explana-tion, even though later disrupted may linger on in the puma as a la-tent but recurrent characteristic.

I make no claim of validity there. But I am convinced that the puma's attitude toward us is ruled by neither hunger nor fear. It is much more influenced by a peculiar respect and an abiding reluc-tance to tangle with us, to cross swords with us, perhaps because in his limited experience with us he has not had time to learn to regard us as either possible prey or potential enemies. He has had only a few thousands of years of experience with us humans, not hundreds of thousands as have many of his Old World relatives — and only a few centuries scaling down in large portions of his range to mere de-cades of experience with us in any real numbers. In most casual meetings with him, unless he as an individual has learned that we regard him as an enemy and therefore should be avoided, we arouse neither his appetite nor his fear-anger — only his ingrained felid curiosity.

Here in North America at least in sharp contrast is his attitude toward the dog, any dog. If a man and a dog encounter him, he is instantly apprehensive — of the dog not the man. A scrawny little yapping pup can put him to flight and tree him and keep him treed for hours. The man can climb into the same tree with him and he will pay scant attention, his apprehension focussed on the pup below. Under such circumstances many pumas have been roped out of trees or snared with wire loops on the ends of sticks. Because of such behavior "cowardly" is the word most frequently used in accounts of him. That word is meaningless applied to any animal but particularly to one who will quite willingly (sometimes successfully) tackle a jaguar or a grizzly bear. Yet it is indisputable that the average North American puma has a strong conviction that anything doglike means serious trouble to be avoided if possible — a conviction fixed in his ancestors to be transmitted to him before human hunting with dogs could have taught it to him.

Again ingenious theories have been offered — that is, by those who know better than to accuse the puma of cowardice. The most ingenious and to me most persuasive puts the responsibility on the wolf. Hunting in packs, the wolf is more than a match for the puma. And the wolf is a very persistent and patient hunter, frequently using relays to wear a victim down. Neither the wolf nor the puma hibernates and in regions of long cold winters food can be scarce. There are known instances of wolf packs pursuing pumas, overtaking and killing them before they could tree or keeping them treed until forced by cold and exhaustion to descend into a desperate fight for life. Who knows how many times through the millenia pumas have met that fate or barely escaped from it or been compelled to abandon their own kills by determined wolf packs? To our pumas anything wolflike, doglike, means deadly danger — and one such doglike creature usually means that others are not far away. It is significant, I think, that on down through Central America and South America where the wolf is unknown, the puma seems normally to have no fear of a dog, will sometimes make a meal of one, and if beset by a pack will rarely show any "cowardly" aversion to making a fight of it.

Nothing of what I have been writing should be taken to mean that I would urge anyone meeting a puma in the wild to try to play games or to strike up a conversation with him. Ninety-nine times out of a hundred the attempt, though probably unsuccessful, would be reasonably safe. But there is always the hundredth chance. The ordinary puma (like the ordinary human) is not dangerous — but the extraordinary puma (like the extraordinary human) can be very dangerous. Though the puma's record as a man mauler and man killer is from our point of view very good, almost as good as the much less numerous jaguar, much lower in score than all of the other large predators, he does have a record. It is a remarkably meagre one, but it exists. He has attacked humans. Drop out of the accounting those in which humans were the initial aggressors and there still remain some forty authenticated cases in which people have been badly mauled without apparent provocation. He has killed humans. Some twenty instances have been well established. Both totals would undoubtedly be higher if more complete records were available from South America. And just a few weeks ago this aspect of the puma was emphasized by the killing of an eight-year-old boy not far from Santa Fe here in New Mexico.

This was a clear case of the hundredth chance. The killer was a rather small female. She was no ordinary puma encountered in her home territory. She had been made extraordinary by hunger, was virtually starving, weighed at least twenty-five pounds less than what would have been normal for her size. She was not hunting, but traveling in an almost straight line through lower country from one mountainous area toward another where presumably she might have better luck at finding food. She came on two boys walking along and jumped one of them. And thereby, of course, signed her own death warrant.

The pumas involved in all such cases on record, I would say, have all been extraordinary. What factors or circumstances made them so? A check through the accumulated accounts indicates that six causes can be listed sometimes with several involved at once. (1)

The puma was rabid. (2) The puma was old and crippled and desperate for any possible prey. (3) The puma was healthy enough but starving. (4) The puma was a female with helpless young nearby and thought they were endangered. (5) The puma mistook movement downwind in brush to indicate another kind of prey and was into an attack before realizing the mistake. (6) The puma attacked out of willfulness or caprice.

That last is the most doubtful of the six, the least susceptible to proof, actually a sort of catchall explanation used to fill in when none of the others can be established. But it must have some validity. The puma is, after all, a carnivore and more particularly a felid carnivore and thus varies in personality — which means a margin of unpredictability.

And so, though the puma's attack record, which has taken more than two centuries to reach its small total, makes him only a minuscule factor in human mortality, he remains a factor. No one meeting him in the wild has the time or the data to determine whether he is an ordinary or extraordinary member of his species. I would suggest no attempt to find out. Meanwhile I reflect that in such a meeting the puma is up against the identical proposition, that of wondering what kind of human he has met — with the odds infinitely more against him. For every attack pumas have made on humans, humans have made thousands upon pumas.

I recently read Theodore Roosevelt's account of just one of his many puma hunts in which he and party killed eleven females (two pregnant) and three males. The last one killed was a big male. "It would be impossible," T.R. wrote, "to wish a better ending to a hunt." My perverse imagination plays with the notion of a puma delivering a neatly lethal neck bite to T.R. as a prize specimen of the human species and regarding that as the perfect ending to a hunt.

In American folklore and the tales of explorers and travelers ever since the first settlements speckled the Atlantic coast pumas have been screaming, shrieking, wailing, but mostly screaming, the sound always blood curdling and terrifying, suggestive of a woman or child in mortal agony. Have pumas really been doing that?

The argument goes on and on. Almost everyone who talks or writes about the puma, nowadays as in the past, feels obliged to discuss the subject. I am no exception. But I propose to discuss it simply to dismiss it. I note that by judicious selection of supposedly authoritative reports I could come up with a yes or no or maybe as desired. On the one hand a list can be compiled of people who insist they have heard a puma scream plus some who claim to have seen as well as heard. On the other hand a matching list can be compiled of people who have lived all their lives in puma country or have hunted pumas for years but who have never heard a puma scream plus others who insist the animal is practically mute. I note further that much pruning could be done on both lists. For example: it is certain that many a reported scream came from the whistle of a steamboat out of sight around a river bend and even more from the throat of the great horned owl who can emit perhaps the most startling sound of any American creature. Moreover any strange sound of any intensity heard in the stillness of night can seem impressively loud and nerve-wracking and he who hears it, aware of the scream tradition, almost automatically ascribes it to a puma — and often amplifies it in the telling about it. Again, since the puma is a solitary hunter and his stalk technique demands silence, there is no particular reason why anyone living in puma territory should ever hear screams — and very good reason why no hunter, almost always aided by extremely vocal dogs, should ever hear a scream. In neither case does the fact of never hearing one rule out the possibility that the puma does scream.

The argument is merely a matter of semantics, a wrangling over words, over what is meant by a blood-curdling terrifying scream. The puma has a proportionally smaller larnyx than the jaguar or the lion or the tiger and his "voice" is therefore less in volume and higher in pitch. He cannot in the same sense emit a "roar"; his equivalent comes closer to being a "scream." He is normally quite quiet, becoming noisy chiefly in the mating season which varies in scheduling according to his environment. In general he can make almost exactly the same sounds as can a domestic cat — only louder. Multiply the volume of the mating caterwaul of a domestic cat and the result can be described by anyone who so wishes as a puma scream — but any blood-curdling or terrifying aspect of it is in the minds of the human hearers. By whatever name it is labeled, it is an affirmation in the felid manner of the common urge of all life to live and reproduce its kind.

To me it is ironic that the puma, potentially among the most formidable of big-game animals, is yet the easiest to kill. Hunting him is usually a strenuous but not much of a sporting sport. It could have more of both qualities if the hunter relied solely on himself and his gun. He rarely does. He uses dogs to do the real work for him. They pick up a scent, start a puma, and chase him, dutifully yipping to proclaim the route. The hunter may get plenty of exercise following because the puma can negotiate very difficult terrain. But the puma is shortwinded, rarely runs any great distance. If he has eaten lately, he will be even less inclined to do much running; he cannot like the wild canids disgorge his stomach contents to unburden himself for a race. He picks a tree and climbs it. The dogs have a fine time on the ground below, proclaiming all the heroic deeds they would perform if that pesky puma would come down — and, of course, signaling the location to the hunter. The latter arrives and can take his time catching his wind, checking his gun, steadying his nerves. The deluded puma is not worrying about him, is concentrating on the dogs. If the hunter waits too long, the puma may leap out to the ground and dash away for another run. This one will likely be shorter and once again the puma is up a tree. The hunter has more exercise and arrives again and from twenty to thirty feet away, closer if he wants, even up in the tree very close to his target, bravely adds another kill to his record.

The puma is not finicky about his menu and will make a meal of almost anything in the meat line he normally encounters in his wild territory. Despite his size he is quite a rodent catcher. But to a considerable extent he preys on members of the deer family.

Long ago the puma and the deer solved the problem of their predator-prey relationship to their mutual advantage. They made it much less of a close interacting relationship than most of us used to believe. The former notion of a puma population increasing or decreasing according to the deer supply and conversely of a deer population doing the same in accord with the number of pumas is now known to be true only in slight degree, barely true at all under normal wilderness conditions.

In a given region under such conditions the puma population will remain fairly stable from year to year regardless of the deer supply. Territoriality is the key. By means of it pumas keep their number in the region virtually constant. On the other hand winter food not predation is the major factor determining the deer population. Their number fluctuates with the availability of food. Puma predation tends to help maintain a balance — for instance, slows down an excessive upward fluctuation pushing past the region's carrying capacity as more weak and diseased individual deer are killed before they starve. During tough winters the presence of pumas actually benefits the deer, who if undisturbed are apt to linger in localized areas called "yards" even though the food supply there is being exhausted, by compelling them to move about more and thus make wider use of what forage is available.

Through thousands of years here they were, widespread across North America, the puma and the deer, both doing quite well. When browsing was poor and the deer supply dwindled, the puma rearranged his menu to include more rodents and lagomorphs and other smaller animals. When browsing was good and the deer increased in number, the puma dined more often on venison — but because he restricted his own population in any specific region he rarely if ever reduced the deer population below the current healthy carrying capacity. Then a whole new factor entered the equation. We humans, just as fond of venison, began in rising numbers to overrun the country.

We constituted a complicated factor deadly for the puma in three ways. For one, we brought with us the age-old "panther" (that should be "pard") complex, the ingrained conviction that anything catlike bigger than a house cat was deadly dangerous and should immediately, if possible, be killed.

For another, in establishing the overblown civilization in which we have so zealously (and perhaps foolishly) enmeshed ourselves, we would steadily and at an accelerating pace destroy the puma's habitats.

For the third, we would inevitably regard the puma (when the wolf was reduced to negligible numbers) as our foremost competitor in predation, competing with us first for food and later for game. The effrontery of the creature! Daring to compete with us! Here is the reaction of a supposed sportsman, Jack O'Connor, as expressed in his book *Hunting in the Southwest* several decades ago — and heavily overweighted in the figure he cited:

> They [pumas] compete with man at every turn and in so doing have signed their death warrant . . . I for one will not regret their passing — if I am still alive when that happens. Each lion killed means at least five hundred more deer for the sportsmen of America, so surely each death should be celebrated and the hunter of lions should have an honorable place among sportsmen.

Mr. O'Connor's wish has come close to fulfillment. By 1971 careful estimates of the puma population of North America could come up with no more than a hundred or so lingering in the Florida Everglades and perhaps five thousand in the whole sweep of the western mountains of Mexico and the United States and Canada.

It is a sad thought to me and should be to all but unrepentant replicas of O'Connor that in the long (becoming now a short) run the puma, the King of American Cats, more a native here than any of us, will remain with us only in a pathetic parody of himself, a captive in zoos, kept as a show.

Order: *Carnivora*
Family: *Procyonidae*

The Persistent Procyonids

"THE MANY species and subspecies of mammals recognized in New Mexico may be grouped under useful species, such as game animals, fur-bearing animals, and destroyers of insects and rodents; or under harmful species, such as predatory animals and rodent pests."

Thus wrote biologist Vernon Bailey in *Mammals of New Mexico,* 1931. He had the general attitude of the time when it seemed relatively easy to rate our fellow creatures in terms of their usefulness to those overlords of creation, ourselves. Nowadays that useful-harmful division is known to be very fuzzy, even in economic terms.

The coyote, for example, is a predator and thus labeled harmful. He is also an efficient destroyer of rodent pests and thus could be labeled useful too. As a predator he quite possibly might nip off a calf or a lamb and arouse a rancher's ire. As a destroyer of rodents he quite certainly would destroy many rodents who, if he were not around, would be busy consuming some of the forage that could fatten the rancher's cattle and sheep. The trouble is that the remains of a coyote-killed calf or lamb are visible and the loss easily figured, while the loss caused by the rodents he would destroy is indirect and difficult to calculate. We overlords usually take the easy way out: kill him as a predator, then extend our killing to the rodent pests by the latest technological means.

Again, there are species who bear fur of market value and thus are labeled useful — and also are efficient destroyers of both insect and rodent pests and useful that way too. But to make use of them in the one way is to cancel their usefulness in the other. We should be grateful that they help us dodge that dilemma by wearing coats of market value only during the cold months. We can let each new crop of them be useful as pest-destroyers during the warm months, then become useful as fur-bearers during the winter. That is, if we are wise enough or they are smart enough to make sure there is always a new crop.

I do not know whether we would be wise enough in regard to the fur-bearers of whom I am considering and contemplating right now. Fortunately we do not have to be. They are smart enough.

Though they belong to the order Carnivora and justify that by ancestry and part of their diets, they have developed their own specialties. Theirs is an exclusive family within the order, the Procyonidae, the raccoon and his allies.

Individualists, these procyonids, despite strong family resemblances. Each at one time and another has been given a whole taxonomic family to himself. Nowadays each is simply given his own genus in the one family and that makes sense. Not merely because of common ancestry, but because they all carry the same banners. Highly efficient tails. Usually ringed in color as if to display a family coat of arms.

Way back when the emerging carnivore order was branching into families, what would be the procyonids chose their particular path. The choice was to abandon a strict carnivorous diet and the chase-or-stalk hunting technique and to adapt to an omnivorous diet and a partly arboreal or tree-climbing life.

They started in that direction during the Oligocene, say some thirty million years ago. By the mid-Miocene, say twenty million years ago, they had completed the change as shown by fossils of a small creature given the genus name *Phlaocyon*. There were varieties of this little fellow and they had the necessary major adaptations: flexible limbs and handlike forepaws for aid in climbing and manipulation of food, and teeth now designed less for ripping and tearing flesh and more for crushing and chewing such other foods as insects and shellfish and vegetable matter. No doubt, though fossils can give only limited evidence of the point, they had banner-waving tails. With rings.

It is likely that all modern procyonids are descendants of this *Phlaocyon*.

He and his were North Americans. As they went on evolving, when the Central America crossing became available, they extended the family range into South America. Meanwhile one branch of the emerging family early invaded Eurasia, using the Bering Strait crossing, and spread across parts of Asia and Europe as far even as England. That was long ago and that line has not fared overwell. Only two lonely genera survive, each with just one species: the red panda of western China and the eastern Himalayas, and the giant panda now found only in a few areas of the high damp bamboo forests of central China.

There have been long and earnest arguments over classification of those two pandas. Since the experts still disagree, I feel free to make my own decisions. I claim the red panda as one of the modern procyonids, an ally at long distance of the raccoon. I refuse, without prejudice, to admit the giant.

The red panda has remained faithful to the family. He looks like a procyonid. And he has kept the family banner in fine condition. Long and well furred. With rings.

The giant panda has gone his own way too long to rate more than remote kinship. I readily agree he is a fascinating animal. But I believe he does not want to be a procyonid. He has tried too hard to be an ursid, a bear. He has the size and the shape of a bear. And, alas, the tail, the almost nontail, of a bear. For me, that clinches the matter. He has spurned the family talisman.

So, though some taxonomists say seven, by my reckoning there are just six procyonids. The red panda lives in far-off China. The other five have never wavered in their allegiance to the Americas.

Two of these Americans are strictly tropical: the kinkajou, who with one species does well in the forests of southern Mexico and Central America and on into South America, and the olingo, who with three species does the same. Both, perhaps because it is a southern fashion, wear fur quite conservative in coloration and have failed to keep the family coat of arms in good condition.

Oh, they have tails, excellent tails. The kinkajou, forgetting or foregoing the rings, has concentrated on another aspect of his. It is prehensile, can be used almost as a fifth limb. He may regard this as distinction enough, since there is only one other member of the carnivore order who can match him in that respect, the binturong of southeast Asia — who belongs to a different family. The olingo seems to be satisfied with size of tail, carrying one large and long in proportion to his small body. No, he is not quite satisfied with that. A close look at him will reveal faint shadowings of rings.

My impression, derived from two I have personally known, is that the kinkajou is an aggravating pushie little beast who delights in using his superb climbing ability to annoy humans and such of their appendages as dogs and cats. Which is just a way of saying that he shares family traits that often turn out that way in captivity. In the wild he may be too busy with more important matters to fit the description.

The olingo I know only from photographs. My tentative verdict is that he is either admirably photogenic or one of the most cuddlesome small mammals evolution has yet produced.

One procyonid in Asia, two living in the American tropics. One and two from six leaves three. These three also inhabit the tropics but are not satisfied with such limited homeland. They hold my major interest because some of all three are fellow citizens of New Mexico. The raccoon of course. And the ringtail. And the coati. Only two other states, Arizona and Texas, can also claim all three.

The foremost among them, again of course, is the raccoon, who leads the family listing with seven species. That number is somewhat misleading, suggests a need for splitting into varieties to hold the family leadership. Not so. The raccoon is remarkably versatile in his one basic type-species and could hold his leadership with that one species.

Five of his other six species have long been isolated on islands off the North American mainland. They are interesting as examples of geographic isolation producing specific distinctions, but they are also limited in range and number. The last of his other six is the crab-eating raccoon, who inhabits the southernmost part of Central America and on into South America.

Enough about the other species. From here on I am writing about that one basic type-species. He is *the* raccoon. Though his banner is, for the family, of only medium length, it sets the family style. Neatly and distinctively ringed, black alternating with grayish buff, ending in a black tip. As if to emphasize leadership, he also wears a black masquerade-style mask.

Being strictly American, he was something new to the early settlers fresh from Europe. Appropriately enough, he acquired his common name from an Indian source, the Algonquin name for him that I have seen variously offered as *arathcone* and *arakun*. He was first named and described in English print by Captain John Smith, who supplied his own rendition: "There is a beast they call *Aroughcun*, much like a badger, but useth to live in trees as squirrels doe." *Aroughcun* into *raccoon* is merely more phonetic juggling plus a dropping of the opening "a."

So he had little difficulty acquiring a common name. When it came to a scientific one, he ran into trouble.

The early Swedish taxonomist, Carolus Linnaeus, had him wrong from the start, thought he had the same yearnings as the giant panda. In 1740, relying on what he had heard and read about this foreign creature, Linnaeus put him down in that basic work, *Systema Naturae,* as a bear, *Ursus cauda elongata,* long-tailed bear. He was right at least on the tail. A few years later, despite having a chance to study a live raccoon, Linnaeus held to that notion. He was sent a raccoon by the then crown prince with orders to examine and describe it. His account, published by the Royal Academy of Science, referred to the beast as an "American bear." Natural enough, perhaps, for a taxonomist who was more of a lumper than a splitter to tuck a new-to-him animal in with others with whom he was already familiar. What may have influenced him most is that the raccoon, like the bear (and us humans), is plantigrade — walks flatfooted on the full soles of his feet.

On through later editions of his basic work Linnaeus insisted on the bear notion, but in the tenth he changed the species label. The genus was still *Ursus* but the species had become *lotor*, the washer. Based on the belief the raccoon washed his food before eating it. Linnaeus was now wrong on both counts.

In 1780 another taxonomist, Gottlieb Conrad Christian Storr of Germany, followed Linnaeus to the extent of keeping the raccoon in the bear family and accepting the *lotor* species label, but he did at least give him the due of a separate genus, making him *Procyon lotor*. Casting about for a genus label, Storr simply plucked one out of the sky. "Procyon" is the name of the Little Dog star. Storr was suggesting that the raccoon is something like a small dog. But not a real one. Still a bear.

That bear notion held out for a long time. As late as 1850 taxonomists were still tossing the raccoon into the bear family. Then a switch. They put him into another family, the Viverridae, the genets and civet cats and mongooses of Africa and Eurasia. That made mild sense. The raccoon and the other procyonids resemble the viverrids much more than they do the ursids, the bears. In fact, the procyonids can be regarded as examples of what is called parallel development in the New World to the vivverids in the Old World.

It was not until along in this century that the raccoon was finally recognized as the ringtailed leader of his own family, Procyonidae, that name based on Storr's genus label for him.

And so, because of the priority rule calling for continued use of whatever binomial name (genus plus species) was given by the first scientist to assign a correct genus rating, the raccoon is and presumably forever will be known in scientific circles as *Procyon lotor*, a designation that suggests he is something like a small dog and he washes his food. Which he is not and he does not.

No doubt, in playing with his scientific name, I am indulging in quibbles. That priority rule has burdened taxonomy with many terms that taken literally would be misleading and sometimes are downright amusing. My biologist son has convinced me the advantages outweigh the disadvantages. Pressed on the point, I would apologize for the quibble about *Procyon*.

I refuse to apologize for the *lotor* quibble.

With that species label Linnaeus (and later Storr) helped establish a myth that has flourished in folklore and books and that was cited to me as gospel about the time I met my first raccoon a half century ago. Later it began to bother me because it suggests that the raccoon is stupid. He is such a cleanly beast, ran the refrain, because he washes his food. He is cleanly all right — as are most animals, especially those furred, when not cooped and corrupted by us humans. But he would be a fool if he regularly washed his food because most of his food does not need washing. And he is definitely not a fool.

Of course he sometimes washes his food — just as we do. When it needs washing and water is available. But food-washing is not a raccoon characteristic, not an automatic action genetically programmed into him or taught him by his elders. He has no interest in justifying a label though given it by so eminent a person as Linnaeus. He has his own reasons for immersing food (as well as other things) in water — and is quite willing to do so in relatively dirty water. That is: food not already well wetted such as that he finds in shallow water along the edges of lakes and streams, which are among his favorite shopping areas. If he comes on food where no water is handy, he will dine without liquid assistance. He will often do so when water is handy.

It is in captivity, cooped in a cage, that he so often seems to wash his food. He will pick up pieces and drop them into his water dish, then feel about to take them out. He is trying to relieve the boredom of confined existence, adding a bit of savor to the food by pretending to find it the way he frequently does in the wild. That is a partial explanation only and there is much more to it.

He is a chewer. He does not gulp food as does the dog and many another carnivore. He likes his food well chewed. Water can soften many foods and moisture aids chewing.

He is a feeler, a handler, a fondler, a manipulator. That is almost an obsession with him. His sense of touch is amazingly developed and he dotes on using it. He will fondle for minutes on end an object whose feel he likes. He not only dotes, he depends on that sense of touch. When fishing in shallow water or reaching for something in a rather inaccessible place, he may seem to be paying scant attention to what he is doing. He is apt to be gazing off into space or scanning his surroundings. His sense of touch is doing the job and needs no visual or other aid. Moreover — and here is the key point — his sense of touch is enhanced when his hands are wet.

He is a dunker, not a washer.

When his hands are wet. His hands. Not forepaws. Not front feet. Those anterior appendages of his are the equivalent of hands. I have what he has not, truly opposable thumbs. But he has in considerable degree what I have only in slight, almost useless, degree, opposability of all his fingers. He can pick up a variety of small objects with one hand, not as I must do by gathering them into my palm and closing fingers around, but by grasping them separately between fingers. His fingernails are not degenerate nuisances like mine. They are stout claws and semiretractile. His hands are surprisingly strong yet capable of flexible deftness and an extraordinary delicacy of touch. They make the forepaws of other carnivores seem crude biological equipment.

His feet (hindpaws, if you wish) like mine are longer than his hands. While their primary function, again like mine, is for locomotion, they have advantages over my inefficient pedal extremities. They are much more flexible, have serviceable toenails helping him include agility in climbing among the methods of locomotion, and their digits, shaming my toes, have some opposability too.

Add to that hand-and-foot versatility one of the highest I.Q.s among carnivores, keenly developed hearing, a wide range of vocal accomplishments, and an insatiable curiosity, and the result is a remarkable beast.

Some time ago I stopped collecting tales of raccoon exploits. There are too many.

His ingenuity in penetrating seemingly impregnable chicken-houses with felonious intent is a timeworn topic. His abilities as pickpocket and burglar are legendary. He enjoys uncorking bottles and sampling their contents. He opens with ease doors not securely locked and considers refrigerators devised for his personal food-finding convenience. He has been known to amble up and help a less well equipped friend, say a dog of his acquaintance, by opening a door for him. Any kind of a fastener that does not include a locked lock will baffle him only briefly. He learns to use the door-bell when making a social call and to operate any kind of signal that has been provided for him when he wants a handout. Who hopes to construct a birdfeeder (except perhaps for hummingbirds) at which he too cannot dine has tackled a tough engineering job. He quickly discovers what faucets are for, easily turns them on, and understands the function of a hose if one is attached.

He finds some television programs worth attention, and switching on a set and tuning for stations a simple procedure. He likes listening to music and has preferences. My favorite among those of him so addicted is one who would tolerate only Bach and Beethoven with a special fondness for Beethoven's Ninth Symphony. He also likes to make his own music, plucking a stringed instrument — and to sing along with his own compositions.

I could go on for pages.

All of which, plus his willingness to return the friendship of worthy humans, makes him a fascinating pet or visiting neighbor — for even-tempered strong-nerved people. The catch is that nothing is sacred to him. He wants to examine, and that thoroughly, anything and everything movable, detachable, fondleable, manipulatable, turn-on-able, openable, tasteable, take-apartable, tear-uppable, etc., etc., etc. The havoc he can create unchecked in a neat household almost passes belief. My favorite tale in that category was reported by Ernest Thompson Seton about a pet of a friend of his. This raccoon broke jail (got out of his cage), explored his jailer's apartment, found a quart bottle of ink, opened it, rejected the liquid as a beverage — and used it to redecorate the room, fingerpainting all available surfaces. The result was trouble for the friend and the amateur decorator, but benefit for Mr. Seton and natural history. This raccoon had demonstrated, especially on the bedclothes, a method of recording animal tracks that Mr. Seton was soon using in his own work.

The above instances refer to him in connection with people and their contrivances. He shows his remarkableness even better in the wild, in the extent of his range and in the persistence with which he maintains himself throughout that range.

The raccoon, that basic type-species, inhabits all of Central America, all of Mexico except an arid piece of Baja California, all of the contiguous United States except the highest regions of the Rocky

Mountains and the lowest of the Mojave and Great Basin Deserts, and large portions of Canada. Oldtime *Phlaocyon* would be proud of him. One single species of a small furred animal conquering most of a major continent.

It is sometimes suggested that we humans have helped him in that conquest by eliminating so many of the larger predators who would willingly make meals of him. Nonsense. He had established most of that range and was around in amazing numbers long before we humans were more than merely foreshadowed in primate development. He did it on his own.

He has continued to do it despite us humans, the most efficient of predators. Many of us have always regarded him, when properly cooked, as eminently edible. For centuries coon hunting has been rated a fine nocturnal sport. In early settlement days his pelt was sometimes used in place of money — payment for services, for example, in so many raccoon skins. Oil extracted from his body was used in dressing leather and his hide was rated excellent for women's shoes. As Peter Kalm noted long ago in *Travels in North America:* "The bone of its male parts is used for a pipe cleaner." Raccoon fur (known in the trade as "Alaska bear" or "Alaska sable") has from the beginning in America been an important item for trappers — and his investigative curiosity has made him readily trapable. His fur has its own value for Davy Crockett headpieces and raccoon coats, but it has also often been passed off, cleverly treated, as that of rarer fur bearers.

As early as the 1740s the Hudson Bay Company was handling 110,000 raccoon pelts in one good month. I note that early in the next century one of the ships dropping anchor by the once famous trading post of Astoria was named the *Raccoon.* But what is really astonishing is that on through the years, while other fur bearers were being trapped close to extinction, the number of raccoon skins taken continued to rise. During the latter half of the last century some 500,000 skins a year reached the fur markets. During the first quarter of this century the average was close to 1,000,000. Because of spoilage and other accidents from trap to market, for a reasonably accurate estimate of animals actually taken such figures must be doubled. In 1948 the Federal Fish and Wildlife Service reported that the annual raccoon market figure sometimes reached 1,500,000. Nowadays, though trapping in general is on the decline, estimates of the annual hunting and trapping toll (not taking into account rising highway kills) are still up near the 1,000,000 mark. Yet the raccoon is still very much with us. His population as a whole is thought to be about the same as it was a century ago. Some experts believe it is increasing. And every now and then evidence is cited that he may be expanding his range in Canada. He is one of, and among the largest of, the animals that are adaptable enough and versatile enough to get along quite well in close proximity to us humans. Just lately I read an account of the surprising numbers of raccoons that can be found living within the metropolitan areas of our biggest cities.

Conservationists have not had to worry about him.

To the raccoon, as to all mammals larger than the small rodents and the shrews and such, man with aid of guns and traps and poisons and dog packs is plainly the worst of all predators. Since he continues to maintain himself against man, I am more than ever convinced that he was quite competent at maintaining himself against all the other predators however numerous before man came along. I like the way naturalist S. G. Goodrich complimented him more than a century ago: "an animal of large resources and marked character."

When developing that sense of touch he did not neglect his other senses. His eyesight is good, adapted to nocturnal food-finding. He has that interesting device of many other mammals which gives them the well-known nighttime eyeshine, a reflecting mirror in the back of each eye. Light not immediately absorbed by the retina is reflected back through for a second chance. His sense of smell too is good. But it is his hearing which may almost match his sense of touch, certainly much better than ours, and his ability to localize sound is much more precise than ours. This is probably his major protective resource. In the wild, whatever he is doing, at the slightest unusual sound he freezes into immobility while his quick brain seeks to interpret it. The instant he knows (even merely suspects) there is danger in it, he is off in search of safety. He is so cautious that when a mild wind rises, apt to deflect sounds, he becomes more wary and slows in his food hunting. If the wind rises more, he will stop feeding and seek a safe hide-out.

Because of that excellent hearing and ability to sort out sounds and because he is sociable with his own kind, he himself is quite vocal. Naturalists trying to describe his vocabulary pile up verbs, assert that he barks, chatters, chuckles, snorts, squalls, growls, hisses, hoots, purrs, screams, screeches, snarls, trills, tremolos, whines, whistles, etc. He does them all and plays variations on each. A gathering when some choice nut or berry is in season will carry on what Sterling North calls a "continuous family conversation." My favorite reference to such a conversation was supplied by Thoreau, who "heard their whinnering at night."

His banner tail is another of his resources. In lush times he uses it to store surplus fat against inevitable lean times. It serves as a brace when he sits on his haunches and thus has hands free for feeling and manipulating. Its most adroit use is for balance when he is climbing or making sudden dodges to evade a pursuer. With its aid he can, like the proverbial cutting horse, spin on a dime and give a nickel change.

When I was much younger and living much further north, rarely even hearing of him in winter, I wondered whether hibernation was one of his resources. Somehow that kind of negative resource, merely a means of enduring part of the year, failed to fit a creature so full of life, so devoted to the investigatable wonders of the world. I was happy to learn that he does not hibernate, instead behaves like the smart fellow he is. He puts on weight in the fall and adjusts his winter schedule to the vagaries of the weather. When the thermometer is too low for outside comfort, he stays indoors and catches up on sleep. Sleep — not the lowered-body-temperature comatose condition of hibernation. When the reading rises, he is out and busy again. He can wait out a cold drop of weeks if necessary, but I suspect that if so he adds another verb to the vocal listing and "grumbles" at so long a wait for the fun of wetted fingers and food fondling.

He may not have to grumble just to himself. Another raccoon or two or more may be with him, companionably sharing body warmth.

Another resource is the scope of his diet. He is omnivorous in the same sense that we humans are, can dine as occasion offers on just about anything edible that he comes on and that suits his fancy at the moment. Linnaeus was impressed by the catholicity of taste shown by his raccoon. "He generally ate whatever came his way: bread, meat, porridge, soups, crayfish and bones — especially the bones of birds, which he chewed as if they were meat. But what he liked best were eggs, almonds, raisins, sugared cakes, sugar, and fruit of every kind." Even earlier, one John Lawson, writing a *History of Carolina,* 1718, had called him "the drunkenest Creature living, if he can get any Liquor that is sweet and strong."

He does, however, draw a dietary line at times. Linnaeus noted that his raccoon "couldn't bear anything with vinegar in it, or sauerkraut, or raw or boiled fish." Since I have similar distastes, I consider that one a high-grade raccoon. His dislike of raw fish strikes me as a good example of the addiction to individuality and idiosyncracy common among raccoons. There have been and are others who rate raw fish delectable. On the other hand, since refrigerators were unknown and iceboxes rare in Linnaeus's time, the fish in question may have been too ripe for a sensitive individual's delectability.

Certainly among the raccoon's major resources is his mate. She is usually somewhat smaller than he, but she is a match for him in every other way and does full duty in maintaining the population quota. In fact, when conditions are favorable, as Thoreau noted had happened when a few raccoons were brought to Martha's Vineyard, she can create a population explosion.

She obligingly matures rapidly and is ready for sex before she is a year old, while he normally has to wait until his second year. She produces young in just over two months from conception and can have as many as seven though three or four is more usual. She takes excellent care of them until they are about a year old — which means that sometimes she has pregnant daughters at home. And she keeps on annually producing, barring accidents, for the ten to twelve years that seem to be the average life span in the wild. She will even adopt orphans and give them proper raccoon upbringing.

The mating habits of them both, him and her, are reminiscent of those of many of us humans, male and female. He has polygamous tendencies; that is, will enjoy sex with several females in sequence as circumstances permit, but is affectionate and attentive to the one involved in the current brief affair. She has monogamous tendencies; that is, will be particular in choosing a sexual partner, often rejecting several suitors before choosing, then is faithful to that one for that season.

As the raccoon is the foremost of the procyonids, the ringtail is the truest aristocrat among them.

He has had at least as much trouble as the raccoon with names, in his case both scientific and common. In both categories the tangles are not yet resolved, so again I make my own decisions.

For quite a while he too was listed as a viverrid. Then for another while and until recently (Bailey was still doing this in 1931) he was given a family all his own, Bassariscidae. Another taxonomic blooper. That name was derived from the Greek meaning "small fox." He is no more closely related to the fox than to various other carnivores and he looks no more like a fox than he does various others — in particular the raccoon. Even now there are naturalists who persist in keeping him in that lonesome family of his own. Those (the majority) who recognize him as a procyonid still keep the small-fox marker fastened to him in the form of his genus name, *Bassariscus*.

Some taxonomists (obviously splitters), while admitting he is a procyonid, insist that some of him belong in a second genus with the peculiar name *Jentinkia*. I follow the lumpers who claim there is just the one genus, *Bassariscus*, with two species, *B. astutus* and *B. sumichrasti*.

B. astutus, while not as much a traveler as his ally the raccoon, yet has a considerable range: most of Mexico, the southwest quarter of the United States, and on up the west coast to Oregon. Now and again I come on suggestions he is expanding that range. *B. sumichrasti* is more provincial, stays far down in the tropical forests of southern Mexico and Central America and is the one the splitters would put in that second genus.

A similar mix-up persists in the matter of a common name in English. Most experts use ringtail. Others use bassarisk or (as did Bailey) the interesting term cacomistle — sometimes spelled cacomixtle. Bassarisk is plainly culled from the genus label and has an unnecessarily exotic flavor. Cacomistle is something else again, is derived from the Aztec name for him and in my opinion should be used, if used at all, to apply only to *B. sumichrasti*. Since this southerner has not had the astuteness of *B. astutus* to explore northward into my New Mexico, I am quite willing to let him have a separate common name, especially one that has had its origin in his part of the continent. He is really a ringtail and should be proud to be, but if I ever meet him in his own territory, I will call him a cacomistle.

Long ago I met *B. astutus* and from the first glimpse found it impossible even to think of tagging him with any of the dozens of other names (in English and Spanish and Indian) that he has had and still has here and there. Those in English make the small-fox notion seem ironic because most of them refer to a different carnivore relative: mountain cat, coon cat, squirrel cat, bandtailed cat, American civet cat. Oldtimers in the west sometimes argued that he was a hybrid, a cross between a cat and a raccoon. He does have a vague resemblance to a cat — as to other small mammals. But if I were a ringtail, I would regard as hopelessly myopic anyone who had a good look at me and thereafter thought of me in connection with anyone but myself.

Ringtail is the right name. Not ringtail cat, as stupid folk who cannot shake the cat habit continue to call him. Ringtail. That name accents precisely what he himself has gloriously accented. His tail is the prize of the procyonid gallery. It is as long as that of the olingo, longer than all the rest of him, body and head combined. It is splendidly, bushily furred. Its rings are sharper, brighter, more numerous, more regal than those of the raccoon. They are alternate dark brown and white and they bring out by contrast the delicate golden buffish tan of his body and lead the eye of the onlooker to the white tips of his forepaws and the marvelous white markings of his pert alert face beneth his huge petal-shaped ears, in particular the white circlets framing his large darkly lustrous eyes. His behavior is in keeping with his appearance. I nominate him for honors as one of the most appealing and graceful and gentlemanly and downright lovely of life's mammalian achievements.

The raccoon, in his impudent manner, accentuates family traits. The ringtail, in his quiet manner, plays them down. He dislikes being excessive or obtrusive. Just as his range is smaller than that of the raccoon, so is he himself. Much smaller. The raccoon tends to be chunky in body like a self-indulgent showoff. The ringtail is leanish, more streamlined, more daintily elegant, and on the average weighs only a fourth as much. He is really a little fellow — except for that magnificent tail. His mate seems to approve the physical scaling down by reducing the six nipples of the female raccoon to her own more modest four.

He is not averse to water, but prefers to use it primarily for drinking, not dunking. He is a capable fondler and manipulator, but his forepaws (I am willing to call them that) cannot really match the raccoon's hands and his sense of touch, dry or wet, is probably less acute. He can be quite vocal with his own scope of sounds and nuances thereof, but except when angry or desperate is never noisy. He has plenty of the procyonid curiosity, but does not indulge this with the obsessive intensity of the raccoon. He is more reserved, more content to do his investigating in natural surroundings and rarely to intrude on human territories. He follows the family custom of being omnivorous, but because of his smaller size and strictly western habitat and usual avoidance of areas where we humans abound and provide garden crops and exotic garbage tidbits, his diet is not as varied as that of the raccoon.

The ringtail too has maintained himself on his own. He has never tried to compete with the raccoon in numbers, but he has kept up his census figures surprisingly well. I say "surprisingly" because it is easy to acquire the impression there are very few of him. Actually there are often more in any locale he approves than people familiar with it ever know. The discovery is sometimes made that he has been living quietly and competently in areas where his presence was before not even suspected. He is so shy and elusive that he is rarely seen. Unlike most of his carnivore relatives, who are most active at dusk and again in the first dim light of morning, he waits until full dark before venturing into the night's adventures and is usually safe home before dawn. He may even decide to stay at home during nights the moon is doing too good a job.

When I first met him I thought at once that he must have a hard time with us humans, that his beautiful coat and superb tail must make him a prize take for hunters and trappers. But his species name, *astutus,* is apt. Very astutely he has developed his own variety of fur. When he is wearing it, alive and well, it is or should be the envy of most other fur-bearers. When he is shot or trapped and it is stripped from him, it soon becomes softish and lax and loses its lustre. As a result, it has had little commercial value with use confined to the trimming on very cheap garments. A trapper concentrating on him would soon go broke. There was one period, fortunately brief, when he suffered from sudden popularity. During World War II when the Office of Price Administration set ceilings on fur prices, the bureaucrats did not know he even existed and he was not on their list. With little or no profit in other pelts, up went the unceilinged price of his. That may be when it acquired the tricky trade name of "California mink" as dealers sought a substitute for the real thing to sell to a gullible public.

Many Americans who have never seen him may have seen his tail, result of an asinine fad that now and then afflicts some people, usually of the male persuasion and prolonged adolescent mentality — that of sporting raccoon tails on automobile antennae or motorcycle handlebars. Occasionally a ringtail tail appears among them. The recognizable difference is that the raccoon tail has five or six light rings while the ringtail tail has a distinct seven. A further check is that the raccoon rings go all the way around while the ringtail rings are not quite complete, have a small gap on the underside.

He is, I submit, at least as *B. astutus,* the most versatile all-around climber of the family — perhaps of all mammals. Some primates and in his own order and family the kinkajou surpass him in forest acrobatics though he is not far behind them, but he has gone beyond them in other forms of agility. Since much of his range is semiarid and some of it actual desert, it is not overburdened with trees and he has to spend a major part of his time on the ground. Moreover, the ground he prefers is rugged — broken rocky terrain replete with canyons and crevices and caves and cliffs. He has adapted himself and learned techniques for negotiating such terrain that should win admiration from bighorn sheep and human mountain climbers.

He has so arranged his rear anatomy that his hind legs can be pointed straight back, parallel to the body axis, and his hind feet can rotate a full 180 degrees. The kinkajou, I believe, has done the same, but has confined uses to tree travel. Because of that rear arrangement, enabling him to anchor himself with his hindfeet and swivel his body in virtually a complete circle, the gymnastics the ringtail can perform on, say, a cliff face almost as well as on a tree trunk are astonishing. He can race headfirst down as well as up, shift direction all around with little trouble, hang and drop from eyelash ledge to ledge, reverse direction right or left on any of them.

In proportion to his streamlined small size and amazing suppleness, his body muscles are surprisingly strong, enabling him to make power leaps with the accuracy of a sharpshooter. That procyonid brain behind his large lustrous eyes has enabled him to develop unusual techniques with those muscles. For example, he is expert at what mountain climbers call chimney stemming. Bracing feet against one side, back against the other, he can rapidly hump his way up or down a narrow crevice. Again, he is equally adept at ricocheting — traversing a wider crevice by bouncing from side to side, reversing position in midair between bounces. To reach a high ledge or the top of a tall rock beyond range of a direct power leap, he can calculate angles like a pool shark and rebound off a nearby perpendicular surface.

In all such activities his tail is a major asset. To see him moving about, especially when he thinks speed is required, shaming a mere squirrel in treetop stunts or soaring from rock to rock, leaping from ledge to ledge, gyrating over a cliff face, with that tail balancing every instant movement, is to realize that with it he has mated beauty and functional efficiency.

The ringtail is as friendly as the raccoon, but as in everything else more reserved in friendship. He is willing to accept you or me or anyone else worthy as an acquaintance, in time as a companion, but we must make the advances and be patient. Gentle and affectionate, he can be one of the finest of pets — except for his usual determination to sleep during the day and be active only at night. Even so, being a gentleman (and his wife a lady) he will adapt to human schedules — as did La Vega, the lady ringtail who had a fourteen-year career at the Desert Museum near Tucson as a photographers' model and television star and as a visitor at schools. A dozen or more youngsters at a time could crowd about her, touching, stroking, examining, and she would give them all amiable response. She would even perch on their heads, wondrous tail curving down, so that they could be photographed wearing live Davy Crockett toppers. Publicity-minded politicians and film people sometimes took advantage of this, but I am sure she preferred posing with children, fellow creatures not yet grown up to the human habit of exploiting anything exploitable.

Beyond doubt the ringtail is one of the most efficient destroyers of Vernon Bailey's pests, insect and rodent. A large portion of his diet consists of noxious insects and an equal or larger of small rodents. In early days in the west he was often called "the miner's friend." A lonely prospector whose cabin was infested by some of the multitudinous varieties of rats and mice might wake one morning to find only a few left and the next morning all of them gone. A ringtail had come along and without disturbing his host's slumber (he has soft pads between his claws) had quietly cleaned them out. The sensible prospector (there were some such) would ponder ways for friendly overtures. If he were lucky, that ringtail (perhaps a pair) would homestead nearby and contribute continued pest control plus occasional more social visits. As Bailey remarked, ringtails "are said to be better mousers than house cats and are certainly more attractive animals." Right. They do not torment their small prey before actual killing or make an elaborate fuss, noisily boasting and showing off the carcasses as do so many house cats.

The male ringtail is not a philanderer like the male raccoon. He is a faithful and responsible spouse. He enjoys his mate's company for more than sex, stays companionably with her during her pregnancy, leaves to find temporary shelter nearby only when the young are born and she is busy nursing them. About weaning time, he moves in again to help feed them and give them basic instruction in hunting — and probably in acrobatics. They grow rapidly and will be ready to strike out on their own in about four months — and he will help send them off and himself remain with her in a companionable twosome. Normally, barring a fatality in the pairing, if you catch a glimpse of him or her, you can be reasonably sure the other is not far away.

Yes, I nominate the ringtail for as quietly gallant and admirable a small beast as we mammals can claim in the whole wide roster of our class.

The coati is the clown of the family. I write that with respect and affection for him, but he does seem to have made himself, in appearance anyway, a caricature of the others — and sometimes his behavior smacks of the same.

He is the largest of the American procyonids. That is a risky statement because it applies only to the male; the female is only about half his size. Moreover he may be outweighed by many a healthy adult raccoon, especially when the latter has fattened up for the winter. But according to available records, the average male coati outweighs the average raccoon by five or more pounds, stands taller, has both a longer body and longer tail. While the raccoon tends to be chunky, the coati tends to be lanky.

His body is longish and flatsided. His hind legs are longer than his front legs and end in feet with longish narrow soles that give the appearance of being turned under. His front feet try to make a match by having longer claws. His usual gait is a shambling loose-jointed amble. You could say that he looks like a raccoon who has stretched himself in various directions, particularly elongating his head, which is pulled out into a narrow mobile always-wriggling snout that turns up on the tip and projects so far that to take a drink he has to immerse part of his face. You could also say that he looks like an oversize ringtail who has let moths work on his body fur and do a real job on his tail. His coat is not exactly fur and not exactly hair, sort of a scraggly mixture of both. His tail, a fair half of his approximate overall four-foot length, is a strong and useful appendage, but it is only sparsely furred or haired or whatever the right word would be and instead of buff or white rings alternating with the family dark brown he has yellowish rings that look as if he had deliberately dirtied them. He ought to be somewhat apologetic about that parody of a procyonid tail. He is not. He flaunts it. He carries it almost straight up like a radio antenna with only the tip bent a bit. When he is roaming through bushes or among rocks, you may see that tail moving about without seeing any of the rest of him. As if to call further attention to it, he has boosted the number of light-colored rings to nine — or sometimes, as Bailey insisted, ten.

Whatever he may be doing he usually does with a solemn serious-ness, but one often belied by the twinkle of seeming humor in his eyes. Watching him, you suspect at times that he knows he is a clown and is proud to be and has deliberately adopted Bob Hope's style of deadpan comedy.

He has not had the trouble of the raccoon and the ringtail with names. His snout has helped. The term *nasua* from the Latin for "nose" has been tagging him ever since Linnaeus started the game and used it for the species label. Linnaeus thought he was a viverrid and made him *Viverra nasua.* The same Storr who gave the raccoon his lasting genus did so for the coati too by the simple expedient of dropping the viverrid notion and using Linnaeus's species name for the genus. *Nasua* the coati became and remains.

Of his three species, *N. narica,* is the one I have met — and the one I regard as the most enterprising of the three, which verdict is based on his willingness to try new territory. He is currently ex-panding his range into the United States.

Down through Mexico and Central America, his traditional range, he has a variety of common names, but the only one I have yet en-countered in English is simply coati. Some naturalists insist on coa-timundi, but that I have been told on good authority is correctly used only for a solitary old male who is either a cantankerous rugged individualist or an outcast from the social troops of his fel-lows.

The lack of English names is undoubtedly due to his relative newness in this country and his still limited range here. He started his immigration about a century ago, crossing the Rio Grande into

Texas, first reported there in 1877. By 1908 he was known to be investigating the southwest corner of New Mexico. By 1924 he was established in the Chiricahua Mountains of adjacent southeast Arizona. Nowadays he has colonized Texas in a broad strip from the Gulf Coast up the Rio Grande to the Big Bend country, expanded his holdings in New Mexico into a big chunk of the southwest quarter, and has pushed in a wide sweep up into Arizona.

He is almost as adept as the raccoon with his forepaws, lacking only the latter's final sensitivity and delicacy of touch. His strong tail and longish hind feet enable him to sit upright (virtually to stand) with forepaws free for action, even such action as skillfully batting insects out of the air.

Like the ringtail he is an excellent climber, lacking only the latter's final fineness of technique and ingenuity. And yet, though he can shame any cat in arboreal acrobatics, he seems to feel safest and most at home on the ground. If startled when aloft in a tree, his immediate impulse is to drop down, literally falling from the lower branches, then to dash for the nearest brush. I have read accounts by observers who had the impression, when surprising a troop of coatis in a tree, that it was "raining" them as they tumbled to the ground and hurried away.

Perhaps that snout has caused him to favor the ground. Or perhaps his fondness for ground-level food hunting has caused him to develop that snout. With its piglike pad of gristle on the tip it enables him to root and burrow and snuffle through debris for insects and grubs and tubers and to poke into holes and crevices after small rodents and lizards — all such items always acceptable because he has the procyonid chewing equipment and digestive system. He even likes to forage under stones, easily tipping up and over those considerably heavier than himself.

He has his full share of procyonid curiosity, but does not make much use of it in regard to us humans and our contrivances and habitations. He is intelligent, remarkably so, virtually a match for the raccoon in that respect. Biologist Marston Bates had plenty of opportunity to observe him while studying the life histories of mosquitoes in the American tropics and concluded that he "must have one of the highest I.Q.s in the animal kingdom." One of the ways he displays that intelligence, I believe, is in his wise preference for sparsely populated areas. His habit (shared by the kinkajou) of traveling in troops could make an invasion of populated areas rather risky. Troops of from ten to twenty of him are common — and I have read reports of as many as two hundred. Invasions in such force would be regarded as definite nuisances and reprisals would be in order.

Traveling in troops, he appears to think there is safety in numbers. At least, when a troop is wandering about, poking into and around and under everything along the way, he behaves as if he regarded himself immune to danger. Often he makes quite a racket as if to advertise his presence. Like the other procyonids he has a considerable vocabulary. He does not have quite the same wide scope of variations on the basic sounds as does the raccoon, so he compensates with an increase in volume. A troop, deadpan clowning, passing remarks back and forth or simply vocalizing comments on the vagaries of existence, can be amazingly noisy. The youngsters step up the seeming babble. They are always roaming off and losing contact with the troop and giving way to hysterical outbursts of shrill squealings that evoke worried coughs and grunts from the mature females. Reunion is finally attained — and in half an hour or so the process is being repeated.

That attitude of unconcern may be merely a part of his clown act. Let any real danger come near and be recognized as such and he shows concentrated concern for safety and survival. It is in his attitude toward the scheduling of activities that he is genuinely unconcerned. He forages for food whenever he happens to be hungry, daytime or nighttime no matter, and does his sleeping the same.

Family responsibilities sit lightly on the male coati for the obvious reason that his half-sized mate carries the full load. He is completely polygamous, in the mating season battles vigorously with other males for as many female favors as he can acquire, then pays scant if any attention to the results born about two and a half months later, usually three to six to a batch. His mate accepts full responsibility, leaving the troop she is with to give birth and nurturing the young on her own until they are able to travel, then taking them with her to join the first troop she can find. Along the way she will add any orphans she encounters to her own brood (one female I have heard about ran up a total of thirteen) and, perhaps joining another matron who has run up another total, may start a whole new troop.

Like his companion procyonids, the coati is responsive to friendly overtures from us humans and can be an amiable pet. That is, if the friendship is forged when he is a youngster. Even when met as an adult, he can be semitamed, but if acquired young and treated right he will become trusting and affectionate. On one of his scientific expeditions William Beebe learned as much from a young coati who made him forget his resolution to have no pets on that particular jaunt. Well aware that scientific objectivity was being forgotten too, Beebe wrote "of the soul of him galloping up and down his slanting log, of the little inner ego, which changed from a wild thing to one who would hurl himself from any height or distance into a lap, confident we would save his neck, welcome him, and waste good time playing the game which he invented, of seeing whether we could touch his cold little snout before he hid it beneath his curved arms."

At times I have wondered whether *Nasua narica*'s continuing range expansion into the United States has been caused, in part at least, by pressures on him below the border. Many people there consider him a protein supplement to ordinary diets and he has been consistently hunted, often with the aid of dogs. Dogs. Plural. He is apt to be more than a match for any single dog. Since we Americans have so efficiently eliminated so many of our larger fellow predators and have as yet no pressing protein pinch in our diets and have not, again as yet, developed coati hunting into a so-called sport, he may think he has found a hospitable new homeland. I wonder too, as we continue to increase our own population and expand our activities, monopolizing ever more of the available environment for our own immediate purposes and pleasures, how long that situation will hold.

It can do us no harm now and again to consider and contemplate of these three procyonids, that friendly rascal raccoon, that gallant little aristocrat the ringtail, and that incorrigible clown the coati. They deserve our admiration because with no deliberate help or protection from us, quite the contrary, they have continued to maintain themselves in this modern world we have been so drastically changing. They have come this far with us on their own. How much farther can they go? Will we continue to think of them as we drive ahead into the even more drastic changes we are now making?

LKPowell

Order: *Chiroptera*
Families: *Phyllostomidae*
         Vespertilionidae
         Molossidae

The Handy Hand-winged

THE VAST MAJORITY of insects fly at some time during their life cycles. Their remote ancestors achieved the ability some three hundred million years ago and most of them have been developing improvements and variations of flight patterns ever since. Nonetheless, in thinking about airborne flight as I am right now, I rule the insects out — in fact try to forget them. They are far too successful at flying, make it seem too simple an accomplishment, and have done so by going in such a completely different evolutionary direction from that of my own ancestors.

We humans, citizens of the province of the animal kingdom known as vertebrata, have every right to be impressed by those of our fellow backboned creatures who have managed airborne flight. That is the one kind of locomotion we ourselves cannot manage in any degree. We can do to some extent just about everything else in that line. We can crawl, walk, run, jump, swim, climb. But we cannot fly. We can concoct machines that fly, but we are simply passengers in them. We control their flying — but we do not do the flying ourselves.

Many people are impressed by what is known of the dinosaurs. I am too. But I am more impressed by the pterosaurs, the flying reptiles of the Mesozoic. They were the first vertebrates to make the breakthrough into the extra dimension of the air, to know what it is not to be earthbound. They invented a new use for their arms and hands — or, if you prefer, their forelegs and forefeet. They lengthened their arms, especially the forearms, then vastly elongated one of the fingers of each hand and by extending webs of skin from their shoulders and arms out to the tips of those elongated fingers they had wings. They must have been rather clumsy and not very expert at flying — but they flew. They were working on the first experimental vertebrate flying model and naturally it was fairly simple and somewhat rickety and could stand improvement.

What we call the age of reptiles came to a close before they had attained much improvement and they joined the parade of extinct forms into the long ago past. But meanwhile some other reptiles, mostly smaller but more inventive, had been developing another flight model and were becoming birds. They discarded the two outside digits on each hand and concentrated on using the remaining three for improved wing structure. They added one of the creative marvels of all time, changed reptilian scales into feathers, and used these for protective body covering, for extension of wing scope, and for a real rudder of a tail. They carried forward the design of a keellike breast for muscle attachment and a lightening of body weight in relation to size and an overall body shape streamlined to reduce air resistance. They avoided a mistake made by the pterosaurs, who were handicapped when on the ground because they had let themselves evolve from reptiles not yet very good at two-legged ground locomotion. The birds had the good sense to evolve from other reptiles already quite adept at such locomotion. This made flight take-off and landing easier for them and enabled them to be equally at home on the ground and in the air. Not content with all that, they got rid of a reptilian disadvantage, cold-bloodedness, and developed one of the general advantages we soon-to-be mammals were working on too, warm-bloodedness.

Our own reptilian ancestors were too intent on becoming mammals to compete with the birds in achieving flight. Perhaps, as anthropocentric people who regard us humans as the goal of the evolutionary process may prefer to think, those ancestors of ours regarded wings as a waste of arms and hands and were looking far into the future to the ultimate triumph of another creative marvel, the human brain. Whatever the reason if any, they forewent attempts at flying and concentrated on such mammalian characteristics as warm-bloodedness, hair-covering, and improved methods of reproduction and child care.

Only after they had achieved these — internal temperature control, fur coats, live birth of babies, mammae feeding of young — did one of the early mammals acquire the notion of airborne flight. It was rather late in the game for that and he had handicaps to overcome. He was committed to four-legged locomotion and to devote any of those legs to wings would make him always awkward when grounded. He was committed to fur not feathers and would have to rely on the old outmoded device of skin-web wings. As an early model mammal he simply was not designed for flight. But he went ahead and achieved it anyway. In fact, he founded the second most successful of mammalian orders, Chiroptera, the "hand-winged," the bats.

Until recently most of us humans were singularly ungrateful. Here was a fellow mammal actually more closely related to us than most of our other mammalian brethren. He was the only one of all us mammals who had managed to uphold the honor of the brotherhood in the matter of flight and had achieved this despite serious obstacles. And how did we treat him? Badly.

The most soothing explanation is that we were led astray by ignorance. Because he was chiefly nocturnal, active only at night, in myth and folklore and art and popular opinion we made him a symbol of darkness and evil, the object of innumerable prejudices and superstitions. Because his flight seemed erratic and unpredictable, we assumed he was stupid and a poor flyer and gave him that essentially derogatory name bat, and from it derived such demeaning adjectives as batty and bat-brained. As if to compound the felony we did almost precisely the opposite in regard to those very distant relatives, the birds, who are alien to us in many ways. Again in myth and folklore and art and popular notions we admired and praised the birds for their flying abilities and often went into raptures to the point of silliness over their manipulation of sounds in supposed songs. Our attitudes toward the two, the bat and the birds, was neatly illustrated by our insistence upon depicting angels as idealized versions of ourselves with bird-wings (minus the bodily distortions actual use of such wings would require) and of depicting demons and devils with forked bat tails and unmistakable bat wings.

With knowledge sometimes comes some wisdom. For quite a while now more and more scientists have been studying the bat. It is now well established that he is as efficient in flight in his way and for his purposes as are the birds in their ways for their purposes. It is now well established that when a bird bursts into song, it is not aiming at beauty and poetic raptures and the pleasing of human ears: it is usually a male warning other males to stay away from a particular piece of real estate or boasting about himself in the hope of attracting a female to share his bed and board. And it is equally well established that in the business of manipulating sounds, in the making and receiving of them, the bat is more ingenious and talented than all of the multitudinous birds except for a very few who have made attempts to imitate his techniques.

Viewpoint is important. And habit of mind. Too many of us, seeing a bat, still automatically think of him in terms of such things as artists' illustrations of Dante's *Inferno* with bat-winged demons flitting about. How would that same bat appear to a mouse? I recall vividly and always with a warm chuckle a drawing that appeared in the *Saturday Evening Post* some years ago: a matron mouse with one of her children beside her, both gazing up at a bevy of bats flying past. "Look," the youngster is saying. "Look, mother, angels."

A full fat volume would be needed for anyone to try to chronicle just the bare outlines of all the weird notions we humans have devised about the bat from as far back in history and prehistory as any records can be found. There is scarcely a body of folklore anywhere around the globe in which he does not figure. The peoples of the Eastern World (and many of the native tribes of the New World) have usually been inclined to regard the bat with favor, considering him variously a bringer of good luck, an embodiment of a guardian spirit, even a sacred being. I find it interesting (and naggingly annoying) that we of the western world have been the chief ones to give him a bad reputation.

Aesop was among those who helped set the fashion. In one of the fables, for instance, the birds and the beasts were at war. When the birds wanted the bat to enlist on their side, he said no, he was a beast. When the beasts wanted to recruit him, he said no, he was a bird. And so, when a peace was arranged, the bat found himself friendless. "Condemned by both sides and acknowledged by neither, the unhappy bat was obliged to skulk away and live in holes and corners, never caring to show his face except in the dusk of twilight."

For the purpose of pointing his moral, the folly of playing both ends against the middle, Aesop took advantage of an argument already old in his time. Was the bat a bird or a beast?

The Old Testament fathers decided one way. In Deuteronomy the bat was listed among the "unclean birds" the faithful were forbidden to eat. Aristotle puzzled some over the problem. "Bats again, if regarded as winged animals, have feet; if regarded as quadrupeds, are without them. So also they have neither the tail of a quadruped nor the tail of a bird . . ." But Aristotle knew that the female bat brings forth live young and suckles them, so he decided the other way. He grouped the bat with the hare and the rat.

Through later centuries a decision was regularly dodged. When described, the bat was simply an "animal." Not rated much of a one at that. In the original *Physiologus* and following medieval Bestiaries he was always "paltry" and "ignoble." I suspect the only reason he was included was that he supplied a chance for another moral. The habit some bats have of roosting together in such thick clusters that latecomers hold to others above them was cited as a kind of dutiful affection — a kind "difficult to find in man."

Early naturalists in these more modern times were trying to classify living things and therefore had to face the problem. As a rule they did what Aesop's unhappy bat did, played both ends against the middle. Since they thought all winged animals should be grouped together, they usually put the bat with the birds — and promptly added at least partial disclaimers, suggesting the bat was neither bird nor beast but something in between. Konrad Gesner, for example, tucked the bat into his volume on birds, then virtually reversed himself: "The bat is the middle-animal between a bird and a mouse so that it may be called a flying-mouse, although it cannot be counted among the birds or else among the mice because it is both shapes."

In his *Synopsis Methodica* of 1693 the English naturalist John Ray was, I believe, the first of the moderns to put the bat down definitely as a beast. A quadruped. A mammal. But how to classify him? Ray was ready for that too. He had a group he called Anomala, anomalies, into which he could pop all puzzlers who refused to fit anywhere else. Into this went the bat along with the hedgehog and the armadillo and various others.

Through the years since the bat in his many incarnations has probably had more taxonomic labels tried out on him than any other mammal. The most interesting treatment to me is that given him by Linnaeus.

In the first edition of *Systema Naturae* in 1735 the bat was listed among the Ferae, a group corresponding roughly with the Carnivora of today. But in the tenth and last edition in 1758 the bat had been honored by being elevated to a place with man himself among the primas, the primates. That decision was based primarily on the fact that with the lone exception of the elephant the bat is the only mammal other than man and the apes whose females suckle the young from a pair of nipples located in the breast region. I suspect Linnaeus was also influenced by the companion fact that has been politely put that the bat's "facilities for elimination and reproduction" closely resemble those of man.

Linnaeus was not far wrong. While the bat deserves what he has had for some time now, his own taxonomic order, it is highly probable that he and we primates had the same ancestral beginnings.

The earliest known fossil bat dates from the Eocene, some fifty-five million years ago, and by that time he was already distinctly himself. His earlier history is still a subject for speculation. But most paleontologists agree he must have evolved from those of the small shrewlike ancestors of all placental mammals who had remained insectivorous and had adopted the arboreal life. In time those tree dwellers developed in two directions, one leading to the early primates, the other to the bat. Through the following ages on to the present the primates radiated out into many life forms, but the bat, having achieved his major characteristics very early, was so successful that he was content with the one fundamental form and merely developed what I call innumerable variations on the single basic bat theme.

With the exception of contemporary man he is the most widely distributed of terrestrial mammals, living on all the continents except Antarctica and on most of the oceanic islands as well. (That he, the lone flyer, could reach those islands while earthbound others could not was one of the items that Darwin stored in memory when voyaging aboard the *Beagle* and noted later in *The Origin of Species*.) The bat can range in body size from a little fellow not much bigger than a shrew to a big fellow as large as a fox. He can vary in dental equipment from twenty teeth to thirty-eight. He is usually an insect eater, but he can also be a meat eater, a blood eater, a fish eater, a fruit eater, even a flower eater. He has the widest array of head and facial peculiarities of all the mammals and can appear (to us humans) all the way from cute and cuddly to hideous and repellent. Trying to keep track of him the taxonomists have listed 4 super-families, 16 families, 19 subfamilies, 180 genera, and at least 980 species. Yet every single one of him anywhere around the world is immediately recognizable as himself, a bat.

The wings do it, of course; are his most noticeable and major identifying feature. He has copied the original wing idea of the Pterosaurs and improved it immensely. Naturally his many families and genera have evolved many variations in details, but all follow the same basic design. Instead of just one elongated finger on each hand for wing support he has gone even the birds one better and uses all four fingers, leaving free only a short sharp-clawed thumb. Since he needs more wing spread in relation to body size than do the proportionately lighter weight (hollow-boned) birds and since he is denied feathers, which could increase the span without increasing the main structure, he has had really to stretch those fingers for length. All are long and the third finger, major in his design, is usually as long as his head and body and legs combined. But because the wing expanse is divided by those digits, it is undoubtedly more flexible and less easily damaged than was that of the pterosaurs.

The wing membrane is thin and quite elastic, fairly strong and almost hairless. It is two-layered, actually extensions of the skin of his back and his belly with no real flesh between, only a small amount of connective tissue interlaced with nerves and blood vessels. It reaches out to those elongated fingers from his lengthened arms and his body and his legs — and many of his species have additional membrane between the legs with the tail incorporated into and bisecting it. Since his legs as well as his arms are committed to his wings, he does not fly as do the birds. It would be more accurate to say that he swims through the air.

I regard those wings as a major evolutionary achievement. Moreover, they are more than wings; they are remarkably sensitive parts with more uses than merely flying. But when I consider and contemplate of the difficulties these wings have imposed on him, I begin to understand why no other mammal has even tried to achieve flight. I note that even the so-called flying squirrel, who apparently made a small start along the same flight path, has to date been content with gliding. To go further would mean giving up what the bat gave up long ago: good ground locomotion and tree-climbing agility.

To have those wings the bat has had to alter completely the structure of his arms and hands. To be able to use those wings he has had to make companion changes in his general anatomy. The act of flying, of moving wings against the air for flight, demands extra support and strength in the chest region. For this he has fused together some of his upper vertebrae, flattened his ribs somewhat, and anchored his shoulder girdle with stout collarbones that reach to his breastbone, which in turn is keeled or ridged for attachment of enlarged muscles. Since his legs are part of his wings, he has had to spread them and rotate them outwards into the plane of the wings — and when he has a tail membrane, to add cartilage spurs called calcars on the inside of his ankle joints to hold and spread that additional wing-surface. His body from the neck down seems disproportionately developed, resembles (in some respects surprisingly so) that of a human who has big burly shoulders and chest outbalancing a narrow waist and narrow hips with thin spraddled legs ending in turned-out feet.

Because of such changes the bat is almost helpless when not in flight, is really a creature of the air. Grounded, he can get about only with a shuffling-hopping gait in what Gilbert White of Selborne described as "a most ridiculous and grotesque manner." I do not think that bothers him much. I might almost say that in order not to be earthbound he has been willing to forsake the earth. Except for a very few very specialized species, the average bat can live out his entire life without ever being on the ground. If he is grounded, that is almost invariably the result of some accident and unless he is seriously injured he can take off again with a clumsy but effective leap using both legs and arms. Even if he lands on his back, he can flip himself over and be on his way again.

When roosting, he is still in the air, not perched like birds but hanging. Since his hands are almost completely devoted to wing structure, he has to depend primarily on his feet for a grip. Since his feet have been swiveled from the usual alignment and his body weight is unbalanced, he cannot grasp and hold upright to a perch as do the birds. He must hook on with his sharp curved claws — which means that the best and safest way to roost is hanging head down. That at least makes take-off into flight a simple matter. He simply lets go and falls free. If instead of hanging somewhere he has crawled into some high crevice or cranny, he simply shuffles to the edge and drops off into flight.

The hanging habit has compelled him, I am sure, to develop some kind of internal control of his blood-flow to offset the gravity pull of blood to his head. He can hang from a slight irregularity of a cave wall or ceiling or tree limb or house rafter — anything in which his claws can obtain a hold — for hours, for days, for weeks, for months, and apparently never suffer even a slight headache. Most of us other mammals would be in trouble if we tried it for as much as half an hour.

His wings impose other seeming handicaps. Their large expanse relative to the rest of him, an expanse unprotected by fur or feathers, presents the twin problems of loss of body moisture and loss of body heat into the surrounding air. The first of these, though he probably requires more water than the rest of us mammals, is not particularly serious. His usual foods are rich in moisture and if necessary he can fly considerable distances to find free water. Even when drinking he shuns the earth. Swooping low, he scoops up a mouthful at a time in flight. It is the heat-loss problem that demands more drastic solutions.

He could solve that one by being content to live in year-round warm climates. But like the stubborn little shrew, who faces the same problem and solves it in a different way, he wants to live just about everywhere on all habitable continents. He does live about everywhere — with one logical limitation. He must have roosting places. And so he ranges only as far north and as far south as trees grow — which is still a long way, thousands of miles from the equator in both directions. Through vast areas of that worldwide ranging the heat-loss problem presses hard on him. Moreover, even those of him who inhabit warm regions create their own heat-loss problem by wanting their sleeptime, the daytime, to resemble the nighttime and many of them spend it in dark caves where temperatures are much lower than those outside. His preference for the dark is so great that those who roost in trees wrap their wings about themselves — some tropical species use palm fronds — to create their own personal small semblances of night.

With solutions to the heat-loss problem he has not been individually inventive. He uses those common to many mammals. But he is adept in the using of them.

His first solution, adequate for regions of only relatively cool temperatures, is his willingness to forego the kind of warm-bloodedness most of us other mammals insist upon. We try to maintain, have to maintain for healthy functioning, a steady, rather high internal temperature. He does not. He can let his body temperature fluctuate to quite an extent according to the air around him and still go about his business without noticeable effect. In cool air, that is, he turns down his internal thermostat. There is a limit below which he cannot turn it and still be active and efficient, but the margin he can use is helpful. Moreover, when he sleeps, say in a quite cool cave, he can turn his thermostat down even further and slumber unworried about too much heat loss. When he wakens, he turns it up and is ready for action.

That solution cannot suffice in regions where winters are winters, where cold temperatures claim long months and scarcity of food doubles the difficulty. In fringe areas like my Southwest where winters are not too long and warm weather not far away, some of him resort to migration. More of him in such areas and all of him in more northern areas hibernate. He does this too with characteristic efficiency.

When the bat hibernates, he hibernates. No restless stirrings and half-awake waiting out of winter cold for him. He drops himself into as profound a "slumber" as any true hibernator anywhere. His body temperature falls until it is barely above his surroundings. His respiration slows until he is breathing only about once every four or five minutes. His pulse dwindles until it is almost imperceptible, which is downright amazing for a creature whose heartbeat rate when he is active is the highest among mammals. His metabolism sinks until his oxygen consumption is only a tiny fraction of normal. Frost can collect on him and he sleeps serenely on.

All the same he has to be selective about winter quarters — which explains why he will be found hibernating in some chambers of a cave and not in others and why he will return to the same proven place year after year and generation after generation. For his purposes the temperature of his hibernation habitat should not be below thirty degrees or above forty degrees Farenheit. I used to think those were absolute limits; that if the temperature fell below thirty he would freeze and if it rose above forty his metabolism would increase while he slept on and he would use up his limited supply of stored fat and literally starve — in either case never wake up again. Now I suspect his internal thermostat can handle such emergencies — if they do not last too long. Recent scientific notes indicate that once he has adjusted himself to deep hibernation he can survive temperatures down almost to zero for relatively short periods. The same temperatures catching him when he is still active would be fatal.

A question to which I hope someday to find an answer is this: how does the bat, sunk into that deep torpor in an inner room of some cave where light never penetrates and where the temperature remains almost constant the year round, know when outside conditions have changed and he can let life really burn in him again? Does he have a cosmic timetable adjusted to his particular region somewhere inside his often curiously ornamented head? Does he have some subtle extra sense denied to or atrophied in the rest of us? Even so, how would such things function when all the rest of him is barely functioning at all?

It occurs to me that his internal-control system must help him solve another problem posed by the expanse of those wings — this not of moisture or heat loss but of heat gain.

His wings have no sweat glands to provide cooling by evaporation. In very warm weather with daytime temperatures above his normal body range, as can occur in the tropics or in desert areas, his wings must be something like blotters, absorbing heat from the surrounding air, tending to induce overheating in him or at least discomfort. If a cave hanger, he can stay in his cool quarters until the usual cooling drop of nighttime or if advisable for a few days until the weather moderates. If a tree hanger, he has no such easy escape. What can he do? He can and apparently does hang quietly in his tree with his thermostat turned well down and his metabolism slowed, thereby conserving energy and establishing a margin for the absorption of heat from the air without that bothering him too much.

All in all I am inclined to think that perhaps his kind of sliding scale warm-bloodedness, helping offset the impact of both cold and heat, is in some respects an improvement on the kind of strict undeviating warm-bloodedness most of us other mammals have — and have to pamper.

All of us mammals are complex creatures. The bat is one of the most complex and primarily so because he has had to devise so many adaptations to meet the problems of aerial existence. Think of this in terms of reproduction and the ensuring of future generations.

Because she spends most of her time in the air, flying or hanging, a female bat would find it virtually impossible to make a nest and raise young in litters. One at a time per year is her usual limit and giving birth to and raising that one is itself a tricky proposition. Moreover, if that young one is to develop into a mature bat with all the specializations required, it needs a relatively long gestation period and subsequent growth period. From the moment of conception, for instance, a house mouse spends three weeks in the womb, is born, and in six more weeks is grown and sexually mature. A bat needs about three months for gestation and at least three more for physical growth and will not be sexually mature until the next year.

In comparison with the reproductive rates of other small mammals, the bat's is very low indeed. It is more than adequate for survival, however, because the bat lifestyle holds predation to a minimum and the bat reproductive lifespan is relatively long. But for bat hibernators the gestation period imposes a new problem, complicates reproductive scheduling. It is both too long and too short.

In regions of real winters if the hibernators waited until spring emergence before mating, birth of the ensuing young would be rather late in the year and the new generation would not have enough time to mature in body and bat winging and bat lore before their own approaching hibernation. On the other hand, if the hibernators mated in the fall before hibernating and the normal course of events coursed on, birth of the young would come due along in late winter with the parents still sunk in torpid slumber. The gestation period is too long for spring mating, too short for fall mating. That hibernators' dilemma has been solved with the usual bat efficiency.

Female bats have developed delayed fertilization. A male can mate in the fall and sink into winter slumber in the assurance that his obliging partner will not let the mating process attain completion until the right time arrives. She does not use the delayed implantation of the fertilized ovum in the womb that some mammals practice. She postpones fertilization itself. She holds his potent sperm within her until, in another of those subtle mysteries of batdom, her internal scheduling gives the signal for fertilization to take place and the new life begin — and begin at the proper time for the birth to be soon after she awakens in the spring.

And so, because of the flexibility available in the business of invoking parenthood, the hibernators can adjust their sexual dalliances to the climatic circumstances of their habitats.

That is all very good, very shrewd and ingenious. But how does a female bat meet the further problem of giving birth up there in the air where she is hanging upside down? Her usual procedure is to reverse position, to reach up and hook on to the roost by her short thumb-claws and let her body swing down to what we would call the right-side-up position. But now the baby as it emerges is in danger of falling to the ground. She cups her tail membrane under her to receive it. If she belongs to a species without much tail membrane, she probably will not reverse position, will remain upside down and simply receive the baby in her wings as if in a basket. Either way she will be aided by the fact that the baby is still held to her by the umbilical cord attached to the placenta within her which itself will not emerge to be discarded for a brief while.

Either way again, her wing membranes will form the baby's first cradle and be the means by which she maneuvers it while she cleans it thoroughly. Once she has done that (and has severed the cord) and has helped it find the nursing site and take firm hold of her body-fur, another young bat is on its way to a possible, even probable, ten to twenty years of airborne life. And even though a bat baby is proportionally large, usually weighing at birth fully a fifth or more as much as does the mother, her versatile wings will enable her to carry it with her on her feeding flights for the few weeks needed for it to become versed enough in bat ways to do its own roosting while waiting for her return.

When handling her baby, a female bat is doing just that: *handling* it, using her hands. She is demonstrating another of the adaptations bats have had to make in order to have wings. Since she has adapted her hands into wings, in a sense sacrificing them to have flight, she has had to adapt her wings for some of the uses of hands. The intricate network of nerves in the wing membranes give them throughout most of their area a highly developed sense of touch. This, coupled with their flexibility, enables a female bat to *handle* her baby. It is also what helps a bat, male or female, be the world's champion eliminator of flying insects.

Though the insect-eating bat (which means most of him) is adept at snatching largish insects out of the air with his jaws, he sometimes uses his wings, especially in nabbing the smaller varieties. He can scoop them out of the air with either wing much as a professional baseball player gathers in a thrown ball with a glove. When one is caught, that network of nerves enables him to feel just where is it and if necessary manipulate it into position to be seized by his mouth. What he cannot do because his fingers are imprisoned in his wing membranes is such grasping work as picking off the wings of certain moths that have no food value or whose taste he dislikes. He has trained his mouth and tongue for that job and can trim such prey much as true southwestern humans can process unshelled piñon nuts in their mouths, eating the nuts while spitting out the shells.

Any ordinary every-night average-kind of insectivorous bat can scoop insects out of the air and manipulate and process them and consign them to his stomach at the rate of one every few seconds — and do this while flying ably along. If he misses one on a first try, he can instantly correct his flight pattern and almost invariably succeed on the second try. No wonder his flight is full of swoops and swerves, of dartings and divings, of leaps and loops, side-slips and somersaults. He is doing his shopping in his supermarket of the air much as a human shopper moves about in a store picking items off the shelves. But he is doing this much more swiftly and adroitly — and combining it with the preparation and eating of the food too. Some of those who appreciate his skill have described him as a wondrously efficient self-guided missile fueled by its own targets. Yet many of us still have the effrontery to call his flight awkward and erratic. It is actually amazingly purposeful, an exhibition of deliberate unmatched aerial acrobatics!

Inevitably and gratefully I associate the bat with the shrew. An unparalleled pair. Our two ablest mammalian allies in the never-ending struggle against our foremost competitors for dominance of the world, the myriad insects. Both are virtually worldwide in distribution, carrying on the good fight in almost all habitable lands. As if for efficiency, to avoid competition of their own, they have divided the battleground between them — and did so far in the past when they were still close relatives. The shrew wages the war on the ground, the bat in the air. Both compensate for small size with multiplicity of troops and voracity of appetite. The shrew devours at least his own weight in insects every day and the bat (he is larger) one third to one half of his weight every night. They are as deadly to our competitors as DDT — and innocent of its deadliness in other directions. They are among the best friends we have in the whole roster of our mammalian fellows. Too long have we repaid them by giving them both bad reputations.

Skill in flight is not enough. Flying insects are skillful too, artful dodgers, elusive and most of them very small — and they have the three-dimensional arena of the air in which to maneuver. The keenest eyesight would barely suffice for their pursuit even in daylight — and the bat pursues them in the dusk and the dark for the very good reason that is when most of them are active and about. How, hunting in the aerial forests of the night, does he track down his prey? How, to widen the inquiry, does he navigate at all when visibility is low or nonexistent, when he is flitting through the tangled branches of close-set trees or threading his way along a twisting passage in the utter darkness of a cave with hundreds of his fellows cluttering the air about him?

As almost everyone knows nowadays, he uses echolocation. Not "radar" as some people insist on saying. That is a man-concocted system using radio waves. The closest of our man-concocted equivalents is "sonar," which uses sound waves traveling in water. The bat uses sound waves traveling in air. He navigates by listening. He hunts with his voice. He sees with his ears.

Ever since the Italian naturalist Spallanzini proved in 1794 that the bat guides his flight by hearing, this particular bat ability has been studied — with special intensity in recent decades — almost as if it were the only aspect of him worth the while. Certainly it is the one that has received the most attention. And still relatively little is really known about it. Oh, the basics are known, the mechanical explanation made, and elaborate tests with species after species have compiled piles of statistics about the sound frequencies used and their modulation and how they are emitted and how received and how the angles of emission and of echo-reception aid in the locating, etc. etc. etc. But the bat has so many varieties of sending and receiving equipment and can use these with such versatility that there is infinitely more to be learned. And how that amazing little computer that is the bat's brain and nervous system operates, turning out instantaneous answers from data that a whole battery of our finest mathematicians using a whole battery of our computers would take hours to analyze and sending out equally instantaneous stimuli to muscular actions, is a mystery I suspect we never will solve.

Consider and contemplate of what an insectivorous bat is doing as he flies along. He is sending out signals from his larynx, some of him through their mouths, some through their noses, in pulses at a rate of many per second. Some use pulses that are frequency modulated, others pulses of constant frequency. The frequencies themselves can range from about 30,000 cycles per second up to around 120,000 — all far above our human hearing whose upper limit is about 15,000. Some of him add a low-frequency component to the pulses that we can detect as a sort of short click. The full sound he emits, if we could hear it, would seem powerful and penetrating, virtually deafening — and could have the same effect on him. So he has a system of blocking his own hearing as he emits a signal, then of freeing it to receive the echo returns. As a further precaution his ears are constructed to avoid sound vibrations transmitted through the bones of his skull. Since he can control both the rate of pulse emissions and the scale of their frequencies, varying both for efficiency according to circumstances and his immediate purposes, he has superbly flexible and sensitive sonar equipment.

He can, for example, return directly in the dark of night to his home roost from a distant hunting foray using solely echolocation and his memory of the route he *heard* on his way out. He can go on hunting in the midst of a night rain, having no trouble in distinguishing between raindrops and tasty insect snacks. Obviously, too, he has no difficulty detecting the echo returns from his personal signals even when countless of his fellows are flying about him emitting their signals. It appears to be impossible to "jam" those signals. Experimenters have tried releasing him in a room full of hanging wires and other obstacles only a bit more widely apart than his wingspread and turning on machines that pour out high-frequency pulses thousands of times stronger than his own. Yet he flies swiftly about, rarely if ever bumping into anything — snatching mosquitoes out of the air.

I have learned that a few of our insect competitors have discovered a way to fool him. No, not to fool him; to escape him. Some moths have developed the ability to hear his signals. Immediately upon hearing, they drop to the ground to wait until he is gone before taking flight again. But I have faith in our mammal friends. Chances are a shrew is down there ready to take over.

I am well aware that many of the items I have been citing about the bat are in some way or ways wrong. He is so numerous and so wide-spread and so various that almost any statement about him is not true for some species. I sometimes think that he, as the only one of us mammals to fly, has felt it incumbent upon him to advance the honor of the brotherhood by exploiting all the possibilities. Or, to approach this from another angle, I could say that the lack of com-petition has enabled him to monopolize the possibilities. In either event he certainly has developed into a surprising array of sizes and shapes and colors, every one of him unmistakably a bat but very many very different in many ways. Such differences show emphat-ically in the variety of head styles.

Whenever I look through my collection of close-up photos of far-scattered species, especially those which show him in profile, I see re-semblances of basic head shapes with those of breeds of dogs. One will remind me of a collie, another of a bulldog, another of a Chi-huahua, another of an Airedale, etc. These differences are proba-bly the result of adaptations of snout and jaws to preferences in diet. But it is his external sonar equipment that gives him his facial pecu-liarities, some attractive, some humorous, some ugly, many down-right fantastic.

Sometimes he has platelike outgrowths on his lower lip. Some-times a band of extra tissue runs between his ears. Sometimes he has fleshy lobes around his mouth producing bulges and folds. Shapes and sizes of ears can vary immensely. Frequently he has an intricate nose-leaf that can follow many patterns surrounding and rising up from his nostrils. Even more frequently he has a tragus, a special lobe at the opening of each ear. These tragi can be weird-looking and occasionally so large one might think they would hamper hearing. No doubt they actually help hearing — though how is still anybody's guess. He takes good care of them and of all his facial features, using his thumb-claws as cleaning tools during the rather prolonged grooming he gives himself whenever he returns to his roost. I infer from the project citations listed in almost every issue of an interesting publication called "Bat Research News" that the puzzle of the why and how of such specialized adornments is keeping a rising number of researchers very busy — and will for a long time to come.

The common names given him in English illustrate the point I am trying to make. All but one of them (flying fox) make him just what he is, the bat, with a descriptive adjective preceding. Once upon a time I started compiling a list of those name-adjectives because they indicate so many of his variations. It is still incomplete, but here it is:

Bamboo (bat)	Flower-faced	Long-eared	Small-footed
Bare-backed	Free	Long-legged	Smoky
Big-eared	Free-tailed	Long-nosed	Spear-nosed
Black-capped	Fringe-lipped	Long-tailed	Spotted
Blossom	Funnel-eared	Long-tongued	Spotted-wing
Borealis	Ghost	Mastiff	Straw-colored
Broad-nosed	Goblin	Mouse-eared	Stripe-faced
Brown	Golden	Mouse-tailed	Sucker-footed
Bulldog	Great-eared	Mustached	Tailless
Butterfly	Groove-lipped	Naked-backed	Tent-making
Club-footed	Hairless	Painted	Thick-thumbed
Cluster	Hairy-footed	Pale	Tomb
Collared	Hairy-legged	Pallid	Trident
Disc-footed	Hairy-winged	Pocketed	Trident-nosed
Disc-winged	Hammer-headed	Red	Tube-nosed
Dog-faced	Harlequin	Round-eared	Vampire
Dome-palate	Hoary	Sac-winged	Velvet-furred
Dusky	Hog-nosed	Shagreen	White-lined
Epauletted	Hollow-faced	Sheath-tailed	White-winged
Evening	Horseshoe	Short-nosed	Wrinkled-faced
False-vampire	Large-eared	Short-tailed	Yellow
Fish-eating	Leaf-chinned	Silver-haired	Yellow-eared
Flat-headed	Leaf-nosed	Simple-nosed	Yellow-shouldered
Flower	Lobe-lipped	Slit-faced	Yellow-winged

Taxonomists divide his order, Chiroptera, into two suborders: Megachiroptera and Microchiroptera, the big hand-winged and the little hand-winged. That seems sensible; on the average the megas are bigger than the micros. But the reverse is true of the suborders themselves. Mega is the small one, comprising just 1 family, 30 genera, 150 species. Micro is the big one with 4 superfamilies, 15 families, 141 genera, at least 830 species.

The megas favor fruit as a major part of their diet and therefore are usually known as frugivorous bats. Probably because, being fruiteaters, they find it convenient when dining to do considerable hanging rightside up, they have a serviceable claw on the tip of the first finger of each hand in addition to the usual thumb-claw. Probably again, being fruit eaters and thus having less need for really fine-tuned sonar, they do not indulge much in special ear lobes and nose-leafs. The multitudinous micros, the insectivorous bats, are the ones who have carried echolocation and delicate devices for it close to perfection.

All of the megas live in the Old World. The micros are world-wide. It follows then that though the New World has no monopoly on them, all of its bats are micros.

Sorting out the world's bats is an impossible task for a late-blooming amateur like me. If I narrow the field down to just the North American bats, I am still in trouble, confronting 8 families, 76 genera, 182 species. If I narrow the field down even more to just the bats of the United States and Canada, I have trimmed the tally to 3 families, 17 genera, 38 species. New Mexico has representatives of all three of those families and a fair share of the genera and species. I am content with them. After all, they cover many of the types and possibilities of batdom.

The Phyllostomidae, the American leaf-nosed bats (the label really means "leaf-mouthed") are a large family living chiefly in the tropics and subtropics. While they have not matched the Old World leaf-nosed in elaborate nose-leafing, they have done fairly well with their own versions. As late as Vernon Bailey's time it was not known that any of them came as far north as the United States. Perhaps only recently have they been extending ranges in this direction. Nowadays two genera, each summing only one species, *Choeronycteris mexicana,* the Mexican long-tongued bat sometimes known as the hog-nosed bat, and *Leptonycteris sanborni,* the long-nosed bat, inhabit New Mexico during the warm months, migrating southward for the winter months.

Both are fair-to-middling insect eaters, but their preference is for flower nectar and pollen — in the gathering of which they provide their proof of the bat's flying abilities. They can hover in front of a flower almost exactly as do hummingbirds. To aid in such feeding they have developed longish narrow snouts and long tongues with tiny projections called papillae on the tips forming a sort of brush. No doubt when pilfering pollen they pay for this by serving as pollenators, particularly for some of our southwestern cacti.

If these two are relative newcomers, moving up here out of Mexico, they are welcome additions to New Mexico batdom and I hope no border patrol ever tries to stop them.

The Vespertilionidae could be called simply the bats because the family name is based on the Latin word for bat. That would not be far wrong because the family is very large and worldwide and includes most of the species with which most of us are familiar. They are sometimes called the simple-nosed bats, which again is not far wrong because only a very few of the many species have any nose adornments and these unspectacular.

Almost all of them are staunch insect eaters, superb flyers in the chase, but at least one genus is known to add to that diet by swooping down over water surfaces to catch small fish. They all have well developed tail membranes, which with most extend to the tip of the tail, which itself extends beyond the outstretched legs. Some are tree dwellers who migrate, more are cave dwellers who hibernate, and there are others who use any hidden dark place such as the attic of an abandoned building, a tunnel, or a deep rock crevice. When coming to roost they usually sweep in head up, hook on by thumb-claws, then toe-claws, and drop down to the upside-down hanging position. If no proper hold is available on a horizontal surface, say a cave roof where they can hang free, they are quite willing to hang on a side wall.

Some seem to be solitary, roosting alone; some associate in pairs; some in small groups; some in big colonies. When the young are being born and will need considerable care, females of the colonial groups band together in maternity wards with the males excluded. I have been told and accept the tale that such females are communally cooperative, will suckle each other's babies as occasion arises.

The genus *Myotis,* the mouse-eared, is the largest and most widely distributed of the family genera — in fact of all the genera of bat-dom, perhaps of all the genera of terrestrial mammals with the exception of the primate genus *Homo.* These mouse-eared are what the majority of us in both the Old and New World mean when we mention bats. They are the most mouselike in size and body appearance and are the ones who have prompted the many common names in many languages that translate as "flying mouse."

In general they are slender in build, well furred, with long tails and tail membranes. Their noses, their snouts, are properly simple for the genus, but they do have fairly long pointed tragi just inside their ears. They are among the most expert of insect snatchers and thus their flight seems particularly erratic and fluttery. Almost invariably they fly with mouths open emitting their supersonic signals. All of them are some shade of brown, paling on the under parts.

There are seventy-one known *Myotis* species around the world, thirteen in the United States and Canada. Nine of the thirteen can be found in New Mexico and my favorite is *M. sublatus,* the small-footed myotis. He is small all over, smallest of the genus and the most solitary, even hibernating alone. He is also, in my opinion, the most attractive, delicately built with long silky fur that is golden brown above, light buff to white below. He is the one any young impressionable mouse might well regard as an angel.

The genus *Pipistrellus* is another of the major genera of the simple-nosed. The name is derived from the Latin for "twittering" or "chirping" and the pipistrelles do plenty of just that when gathered into groups for migration or hibernation according to their environments or when disturbed on their roosts. They are expert and avid insect eaters. In fact, they give the impression of being eager about eating by being among the first of all bats to appear early in the evening, often well before sundown, thus giving observers good chances to watch their swift and darting flight.

There are some fifty species around the world but only two in all of North America. These, however, have quite a range between them and divide it neatly, one inhabiting the eastern portion of the continent, the other the western. *P. hesperus,* the western pipistrelle, is the fellow New Mexican with whom I am acquainted. He is really small, probably the smallest of American bats, quite pale in body coloration. His wing membranes are dark, contrasting sharply with his body and he has sparse fur on about a third of his tail membrane. He is sometimes known as the canyon bat because he is usually found flying not far from the cliffs or canyon walls that seem to be his favorite roosting sites.

The genus *Eptesicus*, the big brown bats, is another of the major simple-nosed genera. The big browns are actually only medium-sized, four to four-and-a-half inches long, big only in relation to such others as the myotises and pipistrelles. Their wingspread of up to thirteen inches helps give them a big appearance when flying.

There are forty-seven known species around the world, but only five in North America — and four of these are confined to lower Mexico and Central America. The remaining one, *E. fuscus,* is *the* American big brown bat. He ranges from far down in the tropics on northward through Mexico, through all of the contiguous United States, and well on into Canada. His favorite habitats seem to be reasonably wooded regions because he likes to dine on the insects that breed there and he appreciates having streams and small lakes nearby because he regards water beetles as very tasty. His usual daytime roost is under loose bark of a dead tree, but he is willing to slip into any hideaway under a house roof, in a barn or other outbuilding, or a cosy rock crevice. The same roost, once chosen, often serves him for hibernating.

His body color varies from pale to dark brown, but his wing membranes are almost black. He usually emerges fairly early, just before sunset, and follows a regular beat among the trees. Since he does most of his hunting relatively close to the ground and in and out among tree branches and has a wide wing spread to worry about, he flies more slowly and deliberately than do the smaller more nimble bats.

The simple-nosed genus *Lasionycteris* is strictly American and has just one species with a name which strikes me as unimaginative in that it could apply to just about all species of bats: *noctivagans,* "night-wanderer." Perhaps I am unfair; this bat has done plenty of wandering as a species. He inhabits the whole of the contiguous

United States and the whole of the southern half of Canada. Here in New Mexico we have large areas of the kind he likes: coniferous forests, particularly ponderosa pine stands spotted with upland parks.

His body fur, which extends to cover a third of his tail membrane, is basically dark brown, but the hair tips are silvery white, giving him his common name, the silver-haired bat. He is medium sized and has rather broad wings that he uses in rather slow beat as he swings in wide circles through his forests. While female bats of any species, like female humans, occasionally give birth to twins, silver-haired females do quite frequently.

The hairy-tailed bats, genus *Lasiurus,* earn their name by having dense fur on their tail membranes. They even add some along the forward edges of their wings. They are often called tree bats not merely because they roost in trees but because of the way they roost. They hang singly or in small groups among clusters of leaves, looking very much like leaves themselves with their coloration adding to the camouflage. They cannot achieve leaf green, but they do well with definite tinges of other leaflike colorings. The three species found in New Mexico, for example, justify their common names of the hoary bat, the red bat, and the yellow bat.

All seven species of the genus are New Worlders and among them have wide ranges from Canada down into South America. They are the only known bats to have more than two young at a time. Rather regularly they have triplets or quadruplets. To take care of these the females break the bat rule and have four mammae. All are strong flyers and those ambitious females have been observed carrying in flight batches of progeny weighing more than themselves.

The genus *Antrozus* is another with just one species, *pallidus,* the American pale or pallid bat, and that one is a true Westerner ranging down the West Coast from Canada to mid Mexico and inland far enough to take in much of the Southwest including all of New Mexico. He is fairly large in bat terms and face-on he has a startled and pugnacious look. Big ears long and broad stand up fanwise and a horseshoe-shaped ridge surrounds his snout close to where this ends in a squarish blunt muzzle.

He is a late emerger, waiting for full dark. His feet are large and strong, aids for one of his habits, that of crawling about on the ground dining on beetles and crickets and grasshoppers and scorpions.

There are four species of simple-nosed bats known as the big-eared, one living far off in the Old World, another a resident of our southeastern states. The remaining two are western Americans and both can be found in New Mexico. They are big-eared all right, very big-eared. *Plecotus townsendii* also has glandular lumps on either side of his snout and is sometimes known as the lump-nosed bat. *Idionycteris phyllotis* has a plain snout but does have two fleshy folds between his big ears that make them seem even bigger and help earn him the sometimes name of the jack-rabbit bat. When roosting both of them roll their ears into spirals before laying them back to rest on their shoulders.

They too are late emergers, waiting for dark, and usually at first swoop about high in the air as if warming up for the night's work before dropping closer to the ground for feeding. While they are expert scoopers, they also often hover close to vegetation, picking insects off the leaves.

The last of my simple-nosed genera, *Euderma,* is another with just one species, *maculatus,* the spotted or pinto bat, but that one is distinctive enough for a fistful of genera.

Back in 1890 the first known specimen was found caught in a wire fence in California. A second was not discovered until 1903. By the early 1960s only about thirty specimens were in all of the zoological collections of the country and he was being described as "America's rarest mammal." In recent years quite a few of him have been found in the southwest and I note that the first ever captured alive was netted here in New Mexico at the Ghost Ranch in the northern part of the state. He is rare, but not as rare as formerly thought. Elusive is a better word — a result of his life style. He prefers privacy, even from other bats, and requires a specialized habitat: arid surroundings for food shopping, moist places in which to roost.

His outstanding physical features are his ears, large, larger even than those of the big-eared, largest of all North American bats, with large tragi. Standing up from a head that seems small in comparison they resemble a pair of swoopingly curved wings. He is undoubtedly one of the most expert of the sonar insect-snatchers, hunting them often in places where they too are rare and elusive. But it is his coloration that sets him apart as a fine and fancy gentleman with perhaps full right to avoid association with more drab bat-brethren. His body is a deep rich reddish brown to black. There is a large pure white spot on each shoulder, another on his rump, and his breast is pure white too. His wings and those great ears are a deep rich glowing pink.

The bat, wrote W. H. Hudson in *The Book of a Naturalist,* is "a very wonderful creature, one of nature's triumphs and masterpieces." And I say that the pinto bat, both efficient and beautiful, is the masterpiece of North American batdom.

The Molossidae are a particularly interesting family. Though they are worldwide in distribution, they have only one fifth as many genera as do the Phyllostomidae, who live only in the Americas. Though they have only about one third as many species as do the Vespertilionidae, they outclass even that extensive family in number of individuals.

They are usually known as the free-tailed bats because they have short interfemoral membranes with a third or more of their tails extending free. Moreover, in most species those membranes are not firmly fastened and can be slid forward some, freeing more of the tail and legs for better locomotion on the ground — or, rather, on the rock surfaces of the crevices into which many of them crawl. As a corollary here they have stouter, better muscled legs than the other bats. Again, their wings are narrower and longer than those of the other bats, which means they can fly faster (*have* to fly faster to sustain flight) but cannot maneuver as well, cannot perform quite the same aerial gymnastics. They are all confirmed insect eaters, chiefly of the largish kinds, especially moths.

The free-tails have structured themselves for speed and endurance in flight. Broader wings and longer tail membranes, as any aerodynamics engineer can explain, would be handicaps, would increase maneuverability, yes, but at the sacrifice of swiftness and durability of flight. So too would be long tapered ears. The free-tails have sculptured their ears to be wider than long with squared tips and able to be laid forward alongside their heads in flight. What if the ears then almost cover the eyes and shut off vision? No problem there for creatures equipped with sonar fully adequate for their purposes. And such ears in such position may help support the relatively heavy heads by providing some lift from the air stream flowing past. Even their fur has been trimmed to a velvety shortness to reduce air resistance.

In regard to wings whether of birds or bats or human aircraft, what is called the camber, the arch or angle they present to the air against which they press in flight, is a major factor in determining the lift provided. The other bats have wings of fairly high camber, which promotes maneuverability but acts as a drag on speed. The free-tails have adjusted their wings for lower camber, which reduces the drag and promotes speed — even demands speed for enough lift to sustain flight, which does not bother them because speed is what they want.

Wings like theirs, beating the air harder and faster than those of the others, need to be tough and well powered. Their membranes have become more leathery and stronger and are better braced by cartilage extensions from the supporting digits. Their flight muscles have extra development — and several specially designed muscles not present in the others have been added.

Wings like theirs, again, cannot serve very well as nets for scooping insects out of the air. The free-tails have to depend more on homing in on their targets and seizing these in their mouths. So some of them have developed large lips that when closed have a wrinkled appearance but when open become quite adequate small scoops.

More so than any of the others the free-tails are the bats of free flight in the open spaces. Little if any foraging for them in and about tree branches and bushes. In bat terms they fly fast and they fly high and they forage far. While their relatives are carrying on the endless war against our insect competitors down closer to the ground, the free-tails are doing their duty too, patrolling with equal efficiency the upper air.

All eight species of the free-tail genus *Eumops*, the mastiff bats, are New Worlders. Six of these can be found in North America, but since they prefer nicely warm climates only one species, *E. perotis,* the greater mastiff bat, has extended his range far enough northward to cross the border into the southwestern United States.

As his name indicates, he is a big fellow — in American bat terms. I believe he is the largest of all bats found in this country and has a wingspread of up to twenty inches. He has large ears joined at their base and projecting forward well past his snout. That snout does have the squarish big-mouthed jowly look of a mastiff. Because of his size and wingspread he needs considerable launching space for flight and usually roosts high enough to have ample space for free fall to become airborne. In my part of the country he is primarily a cliff dweller, though he sometimes roosts up under the overhanging roofs of manmade structures.

The genus *Tadarida* has and needs no other common name than that of the family, simply free-tailed bats. This is the dominant genus, worldwide, and its seventy-four species well outnumber those of all the other family genera combined. Most of them also prefer tropical and subtropical climates and thus the five species found in North America are chiefly Central Americans and Mexicans, but three of the five range far enough north to enter the United States — and all three inhabit New Mexico. *T. brasiliensis,* the Brazilian sometimes also known as the Mexican free-tail, is much the best known, the most studied, and in my part of the country the most famous of all bats.

Here in the Southwest this particular free-tail is usually called the guano bat. He congregates in such vast and closely packed numbers that his droppings accumulate into sizeable deposits of valuable guano. He is also sometimes called the house bat. Though most roost in caves, occasionally a small colony considers an attic or the space between the floors of a building into which they can gain entrance the equivalent of a cave and sets up housekeeping there. Rarely is such tenancy welcome. A free-tail colony is addicted to much squealing and chattering and scratching of claws before settling to sleep — and its homesite acquires an insidious musky odor.

Early one summer evening back in 1901 a young cowboy named Jim White, ambling along on horseback in the foothills of the Guadalupe Mountains of southeastern New Mexico, saw what looked like a whirlwind — or, in the regional phrase, an oversize "dust devil." But it was too dark in color for that. Though it wove and twisted upward, the base of it stayed fixed to one spot. He dismounted and approached, crawling the final stretch, and came to a deep hole in the ground. Out of this were pouring small winged creatures, thousands per minute, spiraling counterclockwise upward. Bats. Brazilian-Mexican-guano-house bats.

Jim White had discovered the Carlsbad Caverns, one of the world's greatest cave systems. Years passed before he could persuade anyone except a younger companion to believe him and help him investigate his find. But in time the caverns and their bats gave him a career, first as foreman of a guano-digging crew, then as a tourist guide, finally as Chief Ranger when President Coolidge established the caverns as a national monument.

On through the years observers mounting into their own thousands and hundreds of thousands came to see what Jim White saw, the evening exodus of the Carlsbad bats — and their equally spectacular return in the first early light of morning.

There they come, high up, out of the far foraging spaces, flying swift and true, gathering into ever larger groups as they approach. Almost straight up from the cavern opening, six hundred to one thousand feet up, they fold their wings and drop into free fall, living projectiles spiraling downward, fast, faster, to plummet in through the opening and sweep unerringly through the increasing dimness and darkness inside to the great domed chamber that dwarfs the Mammoth Cave of Kentucky and is their favorite roosting place. A pulsing throbs the air that is not so much a sound heard as an experience felt. It is not their sonar signals, which our poor ears could never detect. It is the vibrations set up in the air rushing past their wings as they drop out of the sky into the earth.

The nightly exodus and return poses various problems suggesting that *T. brasiliensis* has additional and subtle abilities still beyond our human ability to explain. The myriads spiraling out and upward do not simply scatter to go about foraging in haphazard manner. They seem to follow a relatively orderly procedure as if overall plans for the night's feeding were made in advance. They vary the altitude and the general direction from night to night as if they knew where and at what height the best foraging would be for each foray. They head off and gradually fan out into smaller and smaller groups as if following an organizational pattern — and one roughly equivalent to that they will follow on the return. Since they fly high and forage far, how do they achieve that return? Their eyesight is not particularly good and can be little if any help in the dimness of night — and overcast weather blocking vision of the ground beneath bothers them not at all. They are flying too high for their relatively short-range sonar to guide them. Yet they return as if they had compasses and aerial road maps packed away in their little heads.

When Vernon Bailey was writing his *Animal Life of Carlsbad Caverns* in 1928, he tried to estimate their bat population. He did his figuring in a coldish month and his figure was a mere three million, probably because he did not then know that some of them migrate southward in the colder months. Later studies came up with estimates of nine million for the summer months. Surveys of large caves in Texas yielded even more impressive figures. One cave, logically known as Ney Bat Cave, was found to provide housing for from twenty to thirty million, so many that the last leaving to feed meet the first coming back. Even so, a little arithmetic indicated it would be impossible for all of them to leave during one night. The explanation offered was that the cave's tenants had worked out a stagger system, a rotation of feeding times, so that only about a third emerged each night. If so, that imposed no hardship; experiments have shown that free-tails can get along quite well dining only every third night and if necessary can go without food for even longer periods.

Obviously many bats, the colonial free-tails certainly, have some efficient form of social organization. How do those of Ney Bat Cave, for instance, establish a practicable quota for a night's exodus? How do those whose night it is regulate their leaving so that there is no clogging of the exit by too many trying to crowd through at once? Do they have a status system, one of ranking priorities? Do they have leaders and followers, perhaps in large colonies whole groups within the overall total ranking one above the other? Are the few bats who first rouse as evening approaches, who fly toward the exit and back and repeat, appointed or volunteer scouts whose function is to test the light situation, the state of the deepening dusk, then to summon the others to awaken from the daytime sleeping when conditions are right?

Perhaps the medieval bestiaries were right and the paltry and ignoble little bat has something to teach us. Somehow he has managed to live together in crowded millions more peaceably and more mannerly than we humans have yet been able to do.

afterword

When I wrote the preceding pages about the cave bats of the Southwest I was writing out of memories of years now gone. Time and human tinkering with the environment have robbed them of much of their meaning. Perhaps never again, certainly not for many years to come even if drastic remedial measures are taken, will any of us be able to see and experience what Jim White saw and experienced innumerable times during his career at the Carlsbad Caverns. The millions of bats who once surged spiraling upward into the night's far foraging and returned to plummet in living cascade into the cavern depths have recently dwindled to a mere few tens of thousands. Preliminary findings of biologists conducting a study for the National Park Service indicate that pesticides constitute a major factor. The Carlsbad bats (like most bats) are the last links in a simple food chain: plants to insects to bats. Pesticide residues collect in the bats, are concentrated in the bat-mothers' milk, and are transmitted to the newborn young with deadly results.

Attention is currently focused on the situation at the Caverns because the spectacular flights of the Carlsbad bats have been an important tourist attraction. Sadly I state my belief, virtually my conviction, that if or when surveys are made at other major bat caves, similar results will be reported in degrees varying according to the intensity of pesticide uses in the general foraging areas.

L.K.POWELL

Order: *Edentata*
Family: *Dasypodidae*

Our Lone Edentate

WHY, on my wandering way to becoming an amateur zoologist, do I bother with the nine-banded armadillo? True enough, I have personally made his acquaintance, read considerable about him, met people who say that when properly cooked he is quite tasty, and I once owned a curio made of a part of him. But he remains a stranger, one for whom I can feel scant kinship. He is too different, too much a seeming anachronism out of the long ago past.

Goldsmith put it that the armadillos "seem to be a kind of stranger in nature, creatures taken from some other element, and capriciously thrown to find a precarious subsistence upon land." In these more modern times paleontologist George Gaylord Simpson notes that the armadillos have no "special relatives" anywhere and "resemble nothing else on earth."

More than two centuries ago Linnaeus gave him the genus name *Dasypus* and the descriptive species tag *novemcinctus* and tucked him into an order he called Bestiae. Obviously it was difficult to fit him in and keep him in anywhere. Later taxonomists bounced him from order to order, applied such genus labels as *Cataphractus, Tatusia, Loricatus, Zonoplites, Praopus, Muletia,* and had similar sport with species tags. Nowadays, in accord with the accepted priority rule, he is *Dasypus novemcinctus* again and he and his fellow armadillos have found their place along with those other oddities, the anteaters and the tree sloths, in the small exclusive order Edentata, the toothless.

That is not exactly an accurate label. Anteaters are toothless, but sloths average 20 teeth and armadillos are well supplied. The nine-banded has 32 teeth and the giant armadillo holds the mammalian record with 104. Their teeth, however, are mere rootless blocks of denture with little or no enamel. The polite way to refer to the modern edentates in dental terms is that their teeth have been simplified, reduced, or suppressed. They have been content to concentrate their evolutionary oral energies on development of their tongues. One result, a sort of clearing the way for tongue action, is that none of them have any front teeth.

The first identifiable edentates evolved in North America some sixty-plus million years ago. They kept on here until about twenty-five million years ago and then, never having been very numerous, disappeared. Meanwhile, before the Central American land bridge was submerged, they had spread into South America. During following epochs, while that continent was isolated, they did very well there, becoming important members of the South American fauna. In general they followed two lines of development earlier begun: one leading to ground sloths, tree sloths and anteaters, the other to those experts in armor, the armadillos and what the paleontologists call the glyptodonts.

The ground sloths of the one line and the glyptodonts of the other went in for size. Some ground sloths attained a probable weight of two tons and some glyptodonts achieved body shells five feet and more in length and heavy armored tails ending in spiked knobs. When the Central American land bridge was restored only a few million years ago, both of them, ground sloths and glyptodonts, not only withstood the competition of invading fauna from North America, but made their own invasions northward. On into recent times they were successful inhabitants of North as well as South America. The big ground sloths were still here only a few thousand years ago, a mere tick of geologic time. Sandia and Folsom man here in the Southwest were probably well acquainted with them.

With extinction first of the glyptodonts and then the ground sloths, only yesterday so to speak, the edentates were once again confined primarily to South America with a few species spread into Central America and a few of those edging into southernmost Mexico. Meantime again, very recently, one of the armadillos, my stranger *Dasypus novemcinctus,* started an invasion northward in Mexico up the coastal areas, not over far on the western side but right on up the eastern, the Gulf side. Before long he reached the Rio Grande River — and calmly crossed into Texas.

The earliest printed reference to him there (by Audubon) is dated 1854, range given as "the southern portion of Texas . . . along the lower shores of the Rio Grande." By 1880 and again in 1895 he was being given more range, still in Texas. During the next decades he kept on northward and eastward and by the 1920s had covered most of the eastern half of Texas and was penetrating into Oklahoma and Arkansas and Louisiana. Nowadays he has pushed on northward into Kansas and Missouri and eastward into Mississippi — and with the aid of transportation by man has outposts in Alabama and Georgia and Florida.

While his range expansion north and east was being noted, he must have been quietly scouting in another direction, northwestward up the Pecos River drainage area toward New Mexico. In 1905 his presence there and occasional appearance across the New Mexico line was recognized by biologist Vernon Bailey of the Federal Bureau of Biological Survey. Again, in *Mammals of New Mexico* in 1931, Bailey included him as a "rare resident" of the state. Actually, from the number of his shells that were being and have since been found in that corner of the state, the adjective "rare" could have been dropped.

Then something happened. In 1954 a thorough study of the then current distribution of the nine-banded armadillo published in the "Texas Journal of Science" stated flatly: "Game rangers in New Mexico report no armadillos in that state now."

Oh, well, one might conclude, he had been here, had tested the state, and had abandoned the colonizing project. Not so.

In 1962, as reported in the "Journal of Mammalogy," two armadillos were road-killed in New Mexico. Where? Near Santa Rosa in the central portion of the state. Obviously he was prospecting again — and even further up the Pecos River valley than had those of Bailey's time.

Jump a few more years. In 1970 a friend of mine who gets about in the back country rather regularly stated that he had recently seen several armadillos, even several severals. Where? In the Mosquero area even further up the Pecos valley than where those of 1962 had been road-killed.

But wait. About a year later an inquiry to the State Department of Game and Fish brought a reply that after calls to "the most likely districts" not a single report of an armadillo had been obtained.

Here — then not here. Here again — then not here again. That is a pattern yet again beginning to be repeated. I have an explanation for it.

Edentates in general have never bothered to develop really efficient homeothermy, warm-bloodedness, or the ability to hibernate. Cold weather can be rough on them. *D. novemcinctus* can outlast ordinary cold spells in his usually well-lined burrows, but prolonged severe cold spells can be disastrous for him. I am convinced that he has colonized New Mexico various times and continues to do so and has done fairly well here in favorable periods then been defeated by unfavorable. For example, those of him seen by my friend in 1970 were probably defeated by the prolonged almost unprecedented cold spell of early 1971.

So-o-o-o, with his presence here rendered unpredictable by weather vagaries, I cannot claim that he is at this moment a fellow citizen of my state. I can hazard the suggestion, since at this session of writing the spring of 1974 is sliding into the summer, that he may be started on another colonizing expedition. Certainly I can claim, on the record, that his defeats are apt to be only temporary, that he has shown he wants to be a New Mexican and stubbornly tries to be.

As noted, the modern edentates comprise a small order, just the three families, anteaters and tree sloths and armadillos. The three seem to be utterly dissimilar. They are the last surviving branched-out remnants of a once rather numerous grouping and their taxonomic relationship rests chiefly on a few anatomical items out of the past. I have mentioned that of teeth. Another is the adaptation of their vertebrae by which interlocking knobs strengthen their spinal columns — a development no doubt vital when their ancestors were concentrating on size and armor.

Armadillos, infraorder Cingulata, family Daypodidae, are the most successful of the survivors. They can claim twenty-one living species, while the tree sloths have dwindled to seven and the anteaters to four. They are still specialists in armor, the only mammals so inclined with the possible exception of those far-off exotics the pangolins of Africa and the Far East, who cover themselves with large flattened scales.

As a family the armadillos are low-slung, barrel-bodied, with short sturdy legs, feet equipped with strong digging claws usually four in front and five in back, and long tapering heads ending in piglike muzzles and snouts. Their ears are large and naked. Their armor consists of small checkerlike bones formed in the skin and fused together and includes a head shield, a body shell that hangs down on the sides like a mantle, and articulated rings decreasing in size enclosing the tail. Their legs are covered with hardened scales. In the so-called more primitive types the body shell is composed of movable bands for the full length. In the "more advanced" the body shell has three sections, a solid one in front covering the shoulders, another solid one in back covering the rump, and a set of movable bands between. These last vary in styles: three-banded, six-banded, seven-banded, nine-banded, eleven-banded.

My reference to movable-band styles reminds me that armadillos were among the first New World animals to be included in Old World natural histories. As early as the 1550s Conrad Gesner included in his *Historia Animalium* a surprisingly accurate description — and drawing — of an armadillo. He gave it the curious name of the tatus. Topsell followed him — using the same drawing — and commented that he, Topsell, took it "to be a Brasilian Hedge-hog." I recall counting the bands depicted in that drawing they both used. Precisely and plainly eleven.

The vulnerable parts for all armadillos are the necks and undersides, which have soft skin bearing some scraggly hairs. This, of course, accounts for their best-known defense, that of hiding the soft parts by withdrawing the head between the front legs and snapping the whole body into a ball, which is then encased in armor with only the tail protruding like a handle. One of them even has a notch in the rump plate through which the tail can be withdrawn. Audubon's description was apt: "a small pig in the shell of a turtle."

All of them are prodigious diggers. The largest, the giant armadillo who can attain a body length of four feet and a weight of one hundred pounds, is the most prodigious. Despite his size he can dig himself out of sight through the hardest ground (even asphalt pavement) in about the time it takes me to type this sentence. He plows in first with his front feet, then props himself on his heavy tail so that his hind feet too are free for fast action. With all four feet going strong he resembles one of those endless-chain digging machines that fling dirt in a steady stream. But his major asset in such work is a speciality all his own. He breaks the family rule of four claws on each forefoot and makes that five with the middle one of each set an astounding implement resembling a stout pickaxe — the largest claw not only among mammals but in the entire animal kingdom.

The smallest member of the family, a tiny fellow called the fairy armadillo, is barely five inches long when fully grown. He is pink shelled (the others are a mottled brownish color) and is so elusive in habits and so difficult to keep in captivity that few people have ever seen him. He lives underground like a mole and, being an armadillo, very likely could teach any and all moles lessons in the art of digging.

I suspect that what really intrigues me about *Dasypus novemcinctus* is his current repetition of what the ancestral ground sloths and glyptodonts did long ago, his pioneering up through Mexico into the United States. He is doing it, of course, for precisely the same reason the more recent ancestors of so many of us Americans expanded their range into the United States out of Europe — to find new territory in which to live and make a living.

But why, if he were going to do it, did he not do it long ago? And why has he alone of all his family and order been the one to do it?

My first question may have a simple answer. I offer one — but with no claim it is fully adequate.

Certainly *D. novemcinctus,* even in his subspecies *mexicanus* form, has not recently acquired new evolutionary adaptations giving him greater range capabilities. The expansion I have been discussing has been taking place for little more than a century and is still going on. Yet he himself has been just about exactly what he is now for a very long time. He has not changed. It is the territory into which he has spread that has changed.

Perhaps some slight but sufficient climactic shift a century or two ago triggered his advance northward. True or not, I am reasonably certain that once he had started he found the areas into which he was moving more suitable for him than they had been before. We humans had been working on them.

He likes semidesert or arid, sandy-soiled, fairly open country — open in the sense that it is not forested, that its vegetation is sparse, chiefly scattered thickets and low brush. By deforestation and over-zealous cutting of trees along water courses, by excessive use of underground water lowering water tables, by overgrazing leading to creation of eroded badlands, by all the shortsighted practices with which we have been, in Aldo Leopold's phrase, wrecking landscapes, we have been producing large chunks of the kind of habitat *D. novemcinctus mexicanus* finds attractive.

Even more important, perhaps, we have also been busy eliminating many of the predators (other than ourselves) who could slow down or block his progress.

He is, after all, rather small, not quite three feet long when grown and half of that tall, weighing altogether about ten-to-fifteen pounds. His armor is not much protection because he long ago gave up the size and thickness of plating of his remote ancestors. Moreoever, he can not quite roll up into a fully encased ball as can some others of his family. He does not even try to very often; instead, when threatened, he scoots for one of his burrows (he usually has a number of them) and dives in, wedging himself so that it is difficult to pull him out, meanwhile hoping the disagreeable odor he can emit from anal glands will discourage the attempt.

If no burrow is handy, he may attempt to dig himself out of sight, but more likely he will dash into the nearest thicket. His armor is helpful then, enabling him to plow through faster than can a merely furred pursuer who could easily overtake him in the open. But he has a habit out of the long past that once may have had some sensible purpose but has become a handicap. When startled, he jumps straight up in the air, seeming to need to gather his wits, and thus loses precious time before starting a dash for safety.

I doubt very much that he ever tries the escape technique Goldsmith ascribed to all armadillos, that of snapping into a ball and deliberately rolling over the edge of a nearby precipice "then tumbling down from rock to rock, without the least danger or inconvenience." To do that over any precipice high enough to offer any possibility of escape would be a sure way to commit suicide.

Some people, noting how relatively easy it is to approach quite close to him, have jumped to the conclusion his armor must be so impregnable that, like the skunk and the porcupine, he has small need to fear many potential enemies other than humans. He does seem to go about with what one observer called "fearless oblivion." He is often oblivious all right — but there is nothing fearless about it.

His trouble is that he is not very bright. He is, in mammalian terms, plain stupid. His brain is small, in the evolutionary sense archaic, only the olfactory lobes well developed. Thus his eyesight is poor, his hearing not much better, only his sense of smell acute — and he uses that primarily on ground scents, not wind odors. When he waddles about with snout to the ground in seeming fearless oblivion, he is really concentrating his limited abilities on the search for food, smelling out insects and larvae on and in the ground. Startle him and you will see, following his involuntary spy hop, frantic efforts to find safety fast.

Even his vocal efforts are neglibible. Apparently the only time he emits sound is when frightened and then only wheezing grunts or a kind of buzzing, neither of which seems to indicate more than confusion and fear.

He is so stupid he has not figured out what could, it seems to me, be a possibly effective defense. He has the equipment for it: a head shield, shoulder armor, an extra-strength backbone in the strategic region, and powerful muscles kept in good condition by his burrowing habits.

Lorus and Margery Milne have told, in their *Patterns of Survival,* what happened when they set a nine-banded armadillo down in a closed room. After collecting his sparse wits he snouted about and located the door, correctly estimating it as the thinnest portion of the enclosing walls. He backed off, then dashed straight for it. A few feet from it, without slowing down, he tucked his head between his feet and hurtled into that door. The paneling cracked and splintered and almost gave way. He drew back for another line drive — which no doubt would have gone on through if they had not quickly stopped him.

From that account and somewhat similar reports I conclude that his butting potential is high-power. Adroit use of it might discourage the kind of predators (other than us humans) he is most apt to encounter in his current American travels. They are scarcely in a class with, say, the jaguars of his original territory. But no, he has not been able to calculate beyond the fact that his butting equipment is useful in plunging through thickets. As a result a scrawny coyote that hardly outweighs him can and sometimes does make a meal of him.

I might mention that while he is growing he is particularly vulnerable. Since he cannot shed his armor and develop new as he increases in size, he cannot afford to have it harden into usefulness until he has attained full growth.

All of which is simply to say that determined predators such as pumas and wolves and bobcats and coyotes have no special difficulty making armadillo meals — and that by reducing their numbers we have helped him in his travels.

My other question can, I think, be answered with fair certainty. He alone of his order is able to expand his range northward because he is the least specialized of the edentates. He is more of an opportunist than the others.

The anteaters are completely specialized in diet and are primarily forest dwellers; two of the four species live exclusively arboreal lives high in tropical forests. The same can be said with greater emphasis of the tree sloths. As their name implies, all of the surviving species of sloth are strictly arboreal and eat nothing but leaves and fruit. Some are so finicky they eat nothing but the leaves of certain specific trees. They could not survive outside their forest habitat.

The armadillos other than the nine-banded are also primarily forest dwellers, not as tree climbers but as diet specialists. The giant is a good example. Given his size and equipment, one might think he could more easily than his relatives go where he pleased. But he is tied to his food supply, ants and termites. True enough, Mexico and the United States have shares of both. Not in the quantities he requires. He remains in the deepest of the South American forests where anthills are miniature mountains and termite towns are positive towers.

D. novemcinctus has no aversion to forests and some of his subspecies are quite happy in those of South and Central America and Mexico. But his *mexicanus* is not confined to them, pinned down by dietary and habitat specialization. He dines and with relish on ants and termites, yes, but he is not fussy about tastes. He eats any and all beetles and roaches he can find. He considers spiders and centipedes and millipedes and their numerous relations excellent fare. Fire ants and scorpions and tarantulas simply add spice to his menu. He is not averse to gobbling small reptiles and amphibians. He regards fruits and some vegetables (farmers can testify to this) as acceptable dietary supplements.

With a gustatory appetite like that he has a good chance of filling out his meals wherever he goes. It is not food but weather that will set limits to his range expansion.

Even geographical barriers do not seem to trouble him. He has been found from sea level up to nine thousand feet. Streams, even rivers, big ones like the Mississippi, do not hinder him. His armor makes him so heavy in proportion to size that in water he sinks right down. To cross a stream or small river (southwestern variety) he holds his breath and ambles over on the bottom. If it is too wide for that method, he gulps in air inflating his stomach and intestines until puffed into buoyancy then sculls across to the other shore. Some judicious belching and breaking of wind and he is back to normal.

I maintain that his mate is a real help in his colonizing activities. She keeps the clan sufficiently numerous for expansion projects.

Female anteaters and tree sloths, her companion edentates, produce just one young a year. She is like them in that normally during mating just one ovum is fertilized. But she has found a way to get around that limitation. She practices, the only mammal to do so, what is called polyembryony, the production of a litter from the splitting of one fertilized ovum. On rare occasions a human female tries it — produces identical twins. The female armadillo does much better. Regularly she gives birth to four young and sometimes eight and on very rare occasions twelve. All, of course, identical and of the same sex.

She has also worked out a reproductive scheduling appropriate to northward expansion. She and he mate in late summer, a pattern no doubt established long ago in tropical regions. With a gestation period of four months that could mean the young would be born in midwinter — the time, in more northern regions, they would have least chance of survival. She has so arranged her internal activities that the fertilized ovum is not implanted in her womb and cell division started for about three months after mating, which postpones birth until along in the spring. I would call her a cooperative mate for a northward pioneer.

I also regard it as another evidence of his stupidity that he does not appear to appreciate her. Armadillos may be somewhat social; at least they sometimes go about in small groups and occasionally two or three or four will share the same burrow. But what evidence exists to date suggests that such groups are composed of individuals of the same sex, probably litter mates, polyembryonic identicals. In any event the only time a male seems to pay any particular attention to a female is during the mating season.

She may be at fault there. She has made no attempt to accentuate her femininity. She is the same size and weight as he is, looks exactly like him, acts like him. I would risk a large bet that no mere human, observing a group of armadillos even at close vantage, could tell whether they were male or female or mixed. Perhaps *novem-cinctus* himself can recognize a potential mate only when the urging odor of her estrus helps him. If he meets her at any other time, she may be neither male nor female to him — just another armadillo.

*

If diet alone were considered, the armadillos and the anteaters would be listed in the order Insectivora. Even *D. novemcinctus,* most catholic eater among them, is really a dietary insectivore. His menu is regularly composed from 75 to 90-plus percent of insects and other invertebrates. He is sometimes libeled, accused of destroying the eggs of ground-nesting birds. Once in a long while one of him may acquire a taste for eggs. But ninety-nine times out of a hundred his accusers are misled by circumstantial evidence: seeing him or his tracks near nests in which eggs have been destroyed. Actually he is there or has been there dining on insects attracted to the remnants of eggs destroyed by some previous visitor. Rarely has examination of an armadillo's stomach contents revealed evidence of egg-eating. As a matter of fact, he is a good friend to ground-nesting birds. He keeps down the ants that are a meanace to newly hatched chicks of some birds, of quail in particular. His long sticky tongue, operating through that gap of no front teeth, can sweep up sixty to seventy ants at one swipe. He has been known to consume something like fourteen thousand ants in one meal.

While few of us ever think to give him credit, he and his whole family are good friends to us humans, allies in our constant competition with the insect armies. Looking through a 1943 publication of the Texas Game, Fish and Oyster Commission titled *The Armadillo* in which biologist E. R. Kalmbach reported the results of a thorough study of his subject's supposedly harmful habits, I note that the conclusion reached was that *D. novemcinctus* "has an influence for good." Biologist Vernon Bailey insisted that extension of his range "should be welcomed" and was sufficiently aroused out of a usual scientific calm to claim that the sale of armadillo shells for baskets and curios should be "severely condemned."

Has he been given the welcome Bailey urged? Not exactly, except in an exploitive way by those who regard him as edible or who profit from sales of his shell. He has often been denounced as a pest, chiefly on the basis of the egg-destroying libel and of his occasional rooting in a planted field and of the fact that sometimes an unwary horse with an unwary rider has stepped into one of his burrows and taken a tumble. But now, quite suddenly, he is being warmly welcomed by a sector of our scientists.

The discovery has been made that he can be artificially infected with leprosy — and that he, unlike all other animals similarly tested, can survive the disease long enough for it to reach the late progressive stages. Which means that now he will be *the* test animal in leprosy research. Whole colonies of armadillos will be cooped in laboratories and deliberately subjected to a disease they would not otherwise contract and that will ravage them and doom them to a lingering death.

I refuse to digress into the philosophical and ethical issues involved. This is just another chapter in an old long story. Similar activities involving dozens of other species of animals are going on in laboratories across the country and around the clock and calendar. I will simply state that I recognize the dichotomy inherent in the subject — and inescapably find it in my own mind. When I think in terms of the responsibility for moral choice our capacity for reason should carry with it and in terms of the intricate web of life in which my own species is but one strand, I cannot help feeling ashamed that we humans should treat fellow creatures in that manner. And yet: I know that if I myself, this lone individual sitting here typing, were to contract leprosy, I would be only too glad should a cure, a stoppage, an alleviation of it have been found by research on *D. novemcinctus*.

But consider that discovery about him in another aspect suggested by the tone of the article about it I recently read. Well, well, seemed to be the writer's attitude, what do you know? who would have thought so? "Armadillos do have their uses."

Who *would* have thought so? I doubt whether those who made the discovery thought the chances were good. They and others in the field had experimented with many other animals and failed before they tried the nine-banded armadillo. He was a kind of final forlorn hope.

Here he was, an exotic archaic little beast, one of the few surviving members of an ancient dwindled family, seeming more alien to us humans than almost all others of our fellow mammals. Few people other than zoologists paid much attention to him. The general attitude toward him was indicated by the fact that in portions of his original range he was becoming an endangered species and no one was agitating for protection of him. In this country, to which he came in search of new homesites, he had no importance, not even for sportsmen. And suddenly he has his uses.

He has demonstrated, I say, a practical reason for us to seek never to exterminate, whether deliberately or thoughtlessly, any of our fellow creatures. Who can predict with any pretense of certainty at what time or in what way any one of them may suddenly have his uses?

I said back at the beginning that the nine-banded armadillo remains a stranger to me, one for whom I can feel little kinship. I was wrong — or perhaps I have been teaching myself something. In the process of writing about him, of assembling what I know or think I know about him into some semblance of sequence, I have made a discovery. The faint feeling of kinship has strengthened. He is no longer a stranger. He is a friend and fellow citizen of my state, my country, my world.

L.K.POWELL

Order: *Insectivora*
Family: *Soricidae*

A Brave Little Beast

"From the majority of campers here, as elsewhere, much remains to be desired in camp ethics, especially in guarding the forests from fire and their inhabitants from wanton destruction, in beautifying rather than devastating camp grounds, in guarding streams from pollution."

I could be reading that in many a book or magazine published nowadays. I am not. I am plucking it out of Vernon Bailey's *Life Zones and Crop Zones of New Mexico* — published in 1913.

Has there been any improvement in the more than half a century since? Some. With development of wildlife management programs there is less "wanton destruction" of forest inhabitants. Forest fire-fighting and replanting techniques have advanced. And if a check were made, it might reveal that Bailey's "majority of campers" lacking in camp ethics has become a minority.

But even a minority of campers nowadays can comprise a larger number of individuals than could a majority in Bailey's time. And those minority individuals, without necessarily intending or desiring to be, have become much more efficient at destruction and devastation and pollution.

Which, alas, is an indictment that covers us all, campers or not.

We the people, myself and you and the rest, are no worse, no more ill-intentioned, than we used to be. On the average we may register improvement. But the blunt facts that we keep on multiplying our numbers and increasing our gadgetry and our use of non-natural materials and our fancy packaging of anything and everything make us, will we or nill we, ever more efficient at destruction and devastation and pollution. Simply by being, we all of us partake of the villainy.

We do not even have to try to be villainous. All we have to do is go on living in the style to which we have become accustomed — and go on increasing our numbers. Inevitably and inexorably we are compelled to monopolize ever more of this planet's limited capacity to support life, crowding toward extinction ever more of our fellow creatures.

And, perhaps, ultimately eliminating even ourselves.

Someone should do something. I offer the SSS, the Schaefer Sizing Solution.

Inspiration came while I was reading Julian Huxley's *Evolution in Action*. He was making the point that nowadays, in regard to evolution, "we know that artificial selection can be more effective and can get results much more quickly than natural selection." He cited the horse, which man by artificial selection has pushed in strength and speed well beyond what nature had accomplished by natural selection. Then he made the jump, applied the principle to people: "On theoretical grounds, we could certainly breed up a number of specialized human types if we set our minds to it."

Why not set our minds to it? Aim at a human type specialized in one particular respect. That of size. Set a goal for ourselves of, say, one half the present norm.

Individual size and strength and speed are no longer of any real importance to us except in certain sports and even in these sports differences are purely relative. Machines do our major work. Overall body size has scant if any relation to intelligence. We could all be much smaller and be just as smart or stupid, as lazy or ambitious, as happy or unhappy, as we are right now. And our individual impacts on the finite resources and limited area of this spaceship Earth would be proportionately smaller.

Think about this with your oversize brain in your oversize body.

If we began sizing ourselves down, the per capita reduction in need for materials and space might offset the increase in population that we obviously have difficulty in merely slowing down. By time we had reached the half-size goal with our gadgets and appurtenances similarly reduced in size, the human population of the world could have doubled without any greater strain on resources and space than we have right now.

Then, if necessary, we could renew the goal: one half of the one half.

J. Huxley only triggered inspiration. Background came from my current interest in the subject of size, in turn brought on by considering and contemplating of the smallest of mammals, the ones I regard as my earliest true mammalian ancestors. The soricines. Family name Soricidae from the Latin *soricis,* shrew-mice.

They are not mice, though the Romans (and before that the Greeks) thought they were at least close relatives of mice. They are something quite different. They are shrews.

True enough, when seen in quick glance (the only way they usually are seen) most species of shrew look like some species of mice. Small furred four-legged beasts with slick tails. A less quick glance may catch the fact that these beasts are much smaller than mice and have heads proportionately longer, narrowing to somewhat flexible snouts. There are scores of other differences, both external and internal, enough to fill pages. I will mention a few as I ramble on.

Mice are rodents, younger creatures in an evolutionary sense. Shrews are insectivores, order Insectivora. They are the only members of that order living in my New Mexico. That is not surprising, since in all of the United States there are only two families of insectivores, the shrews and the moles, the latter much less versatile in genera and species and much less energetic in extent of range, most of them staying conservatively in the eastern states. No doubt I am prejudiced, but I do not consider the absence of moles any great gap in New Mexican wildlife.

As a family the shrews are the smallest of mammals. Only a few species, even when fully mature, weigh up to one ounce, the wide majority much less and averaging in length no more than three to four inches including longish tail. The smallest of these smallest is Savi's pygmy shrew, native of the eastern Mediterranean area, who has to stretch to be all of one and a half inches long with nearly two thirds of that tail and who weighs less than a worn dime.

Even those shrews who seem to have some interest in size do not attempt much. The largest is the goliath shrew of Africa, who sometimes attains a body length of six inches with four more of tail. Perhaps he has achieved that much by association with us overgrown humans. He likes to live about or in African-style houses and other shelters, feeding on those frequent companions of humans, cockroaches.

Plainly smallness is no handicap for survival. Not even for mammals, who of necessity have complicated internal mechanisms. Paleontologists and evolutionists generally agree that the common ancestors of all us placental mammals were tiny creatures very similar to modern shrews. The inference is that, while many of those early shrewlike mammals busied themselves founding future mammalian orders, some of them were content to remain simply shrewlike. The further inference is that what we nowadays call shrews have been around without major changes for a long long time — are probably the oldest of placental mammals.

As another item in probable ancestry I note that all shrews, just as they did scores of millions of years ago, have five digits on each foot, unlike the rodential mice who get along with four. To look closely at those tiny five-digited feet is to see a far-off foreshadowing of the fingers with which I am typing this and the toes I am wriggling inside my shoes.

Plainly again, smallness is no handicap to successful life on this aging earth. Not even for mammals. The three mammalian orders (other than our own) most successful in number of species and number of individuals are also the three whose individuals average out the smallest: rodents and bats and insectivores.

The smallest in individual size of the three, the shrews, comprise by far the largest division of the insectivore order. Their family can claim 3 subfamilies, 21 genera, 265 species. According to specific tastes they are at home in water, on deserts, throughout grasslands and brushlands and forests, and from sea level up toward the Arctic-Alpine of the taller mountains.

They have their enemies, of course: the raptor birds, some snakes, and such larger mammals as weasels, foxes, bobcats, coyotes and the equivalents of these in other continents. But they have their defenses. Their very smallness is one, enabling them to duck into hiding almost anywhere. They are so quick in action too that they can be difficult to catch. Experiments have shown that an average shrew can scamper at a speed of two miles an hour — which does not seem like much to bigsters like us humans, but for beasts so tiny is the equivalent of well over 100 miles an hour for us. Coupled with superb supple agility, that relative speed can baffle many a predator.

They have, moreover, a musk gland in each flank that can emit an unpleasant odor and apparently can give an unappetizing taste to their small carcasses. Observation indicates that only such enemies as hawks and owls and snakes, who seem to be indifferent to taste in food, rather regularly try to dine on shrews, while their mammalian predator cousins go after them only when very hungry.

All American shrews belong to the subfamily Soricinae, the red-toothed shrews. Their teeth are not really red, merely reddish brown toward the tips, but that Amerindianish characteristic helps set them apart from the majority of the shrews of the rest of the world who are white-toothed.

Our patriotic red-toothed can not match the foreign white-toothed in number of genera and species, but they do rather well: five genera, sixty species.

Every section of the United States has its shrews, but there are geographical distinctions. The long-tailed (genus *Sorex*) must be the happiest here; they have the most species and these cover the whole country. The short-tailed (genus *Blarina*) and the small-eared (genus *Cryptotis*) are more provincial, inhabit only the eastern half of the country. The pygmy (genus *Microsorex*) prefer a coldish climate and live in the northern states. The desert (genus *Notiosorex*) have opted for the arid life, restrict themselves to the Southwest.

If you have followed me through the above paragraph, you may have noted that 2 of the 5 genera include New Mexico in their ranges. We have here 5 of the 31 species of long-tailed and 1 of the 2 species of desert.

S. vagrans, the vagrant shrew, is the most numerous. He is reddish brown above and gray below and he likes mountain areas, inhabiting them here from the lower levels all the way up to the real heights. *S. merriami,* Merriam's shrew, is the shyest and most unobtrusive. He is grayish above and whitish below and spends his time in open, rather arid back-country where he is not apt to encounter many overgrown humans. *S. cinereus,* the dusky shrew, is really a far Northerner, ranging all over Alaska and Canada and the northern tier of our states from coast to coast, but he has followed the Rocky Mountains down into my state too and has taken over the Canadian Zone of the Sangre de Cristo and Jemez and San Juan Mountains. He holds to the pattern of brownish above and grayish below, but in winter his back becomes sooty, almost black, and he can then boast of being tricolor. Biologist Bailey put him down as the smallest of New Mexican shrews. He is not. *S. nanus,* the dwarf shrew, is the smallest, identified here since Bailey's time in the Sandia Mountains (which I can see out my study window) and in the Las Vegas area. He can compete with the pygmies of other parts of the world in smallness.

S. palustris, the mountain water shrew, is the sartorial dandy among our shrews. His fur is dense and velvety, brownish black above and silvery white below, and in winter acquires a glossy sheen. Most of his time is spent in streams and ponds where, aided by fringes of stiff hairs on his hindlegs and underpart of his tail, he is a superb swimmer, submerging to catch minnows with seeming ease or skimming over the surface after insects and waterbugs. He too is primarily a far Northerner, but since he too inhabits the whole long sweep of the Rockies, he enjoys life in New Mexican mountains.

N. crawfordi, Crawford's desert shrew, is either quite rare or remarkably elusive even for a shrew. Few specimens have ever been collected and little is really known about him. He is gray above and the same though lightening below and has a shorter tail than his *Sorex* cousins, thinner fur and larger ears — which last prompted Bailey to call him the eared shrew.

Enough of such taxonomic juggling. In rambling on I can forget genera and species and subspecies. A shrew is a shrew is a shrew. The family is remarkably similar throughout in habits and particularly in temperament. With good reason too.

Some of my favorite reading is in a bulging two-volume work by S. G. Goodrich published in 1861: *Illustrated Natural History of the Animal Kingdom*. He was friendly toward shrews — and annoyed that, "among the ancients, the shrew-mouse had a very bad reputation." Right. And not just among the ancients.

Aristotle claimed that "the bite of the shrew-mouse is dangerous to horses and other draught animals as well; it is followed by boils." He believed "the bite is all the more dangerous if the mouse be pregnant when she bites, for the boils then burst, but do not burst otherwise." Pliny and Agricola and others echoed him through the centuries. By time Topsell compiled his compendium of fact and fiction in 1607 the shrew had become a fearsome creature with "four rows of teeth, two beneath, and two above" rendering its bite doubly deadly. To Topsell the smallest of his mammalian fellows was "a ravening Beast" one that "beareth a cruel minde, desiring to hurt anything, neither is there any creature that it loveth, or it loveth him." To Topsell it was fortunate that "in old age they become utterly blinde by the Providence of God, abridging their malice, that when their teeth are grown to be most sharp, and they most full of poison, then they should not see whom nor where to vent it."

Again: to Topsell it seemed logical that a creature so violent and virulent should possess, when properly processed, powerful medicinal properties. He collected whole folio pages of prescriptions for use of various parts of the shrew in the preparation of medicines certain to cure numerous human ailments. It bothered him not at all that the methods of preparation involved just as much human malice and cruelty as he ascribed to the shrew. For example: for certain medicines, the shrew had to be caught, hung up alive until it starved to death and became stiff before used. For other remedies requiring the tail, considered particularly potent, that tail had to be cut from a living shrew or it would lose its properties.

Quite a while later even the gentle Gilbert White of Selborne was reporting the belief that "a shrew-mouse is of so baneful and deleterious a nature, that wherever it creeps over a beast, be it horse, cow, or sheep, the suffering animal is afflicted with cruel anguish, and threatened with the loss of the limb." What was the supposed cure? To have a "shrew-ash" handy and apply twigs or branches from this to the afflicted limbs. And what was a shrew-ash? An ash tree into which "a deep hole was bored with an augur" into which a live shrew-mouse was stuffed, then the hole was plugged.

In modern times criticism of the shrew's character has bristled even more with adjectives like aggressive, violent, ferocious, bloodthirsty. As early as Shakespeare's time (he certainly helped this along) the label "shrew" was already commonly applied to a woman with a scolding, nagging disposition. The word still has that meaning in current dictionaries. For my part I agree with Goodrich, who found it "difficult to account for such widely extended prejudices" in regard to "this graceful and harmless little animal."

Basis does exist for Aristotle's "dangerous" accusation. Some species of shrew (not the western long-tailed) are venomous, the only placental mammals so endowed. I have to limit that to placentals because the male of the curious half-mammal, the platypus, has poison glands from which he can eject venom through hollow spurs on his hind legs. Shrews poison by biting, the venom coming from specialized salivary glands. The immediate effect is to slow down the heart action and breathing of the creatures bitten. But shrews are so small that their bites can affect with serious consequences only the small prey they normally pursue. A shrew bite could barely bother one of us and people who have handled them (even made pets of them) say they show little inclination to try to bite humans. A healthy horse, if actually bitten (I doubt if a shrew could bite through horsehide), would probably not know it had been. My belief is that any bad effects observed back in Aristotle's time and on up through Topsell's and White's were the result of bacterial infections through the tiny punctures made by the shrews, not from the infinitesimal amounts of possible venom.

The ferocity charge might seem to have more real basis in fact. No doubt about it, the shrew is a mighty hunter, a deadly fighter, a confirmed killer. In shrew-size terms. Many naturalists have written variations on the theme that other creatures, we humans among them, are lucky that shrews are not the size of tigers because they would then be the deadliest of all beasts. Nonsense. If shrews were the size even of small house cats, they would no longer be shrewlike in temperament. The basis for it would have been removed. They are not ferocious. They are simply hungry. Very hungry. Virtually all of the time.

Smallness does present a penalty — when carried to the mammalian extreme of the shrew.

There are natural size limitations on land mammals in both directions — in largeness and in smallness. The crucial factor for largeness is bone structure, for smallness is surface area.

A land mammal can stand up on its feet and move about because it has an internal skeleton with important key bones to support its weight. If it should double in size throughout, it would square the carrying capacity of those bones — but would cube its volume, its weight. That is: as size increases, overall weight rises much more rapidly than does the skeletal supporting capacity. Which means, in actual evolutionary development, the evolving mammal must increase the proportionate size of the supporting bones faster than it does the overall body size. All manner of variations are possible, of course, such as improving bone composition or increasing only selected portions of the overall anatomy. The giraffe, for example, embodies a batch of variations. But the basic principle remains: Beyond an optimum ratio between bone and flesh for each type of body structure (say, that reached by the pronghorn for hoofed mammals) the larger the beast the more cumbersome it becomes because of the necessity for increased bone structure. A mature African elephant weighing about ten tons is probably close to the largeness limit.

Some of the dinosaurs (they were *not* mammals) in their time grew well past the limit. They did it by living out their lives in swamps and marshes where water and watery ooze helped support their outsize weight — and in so doing helped along their own extinction as their watery habitats dried away in the swing of geologic time. Whales (they *are* mammals), having chosen a completely watery habitat, though they have little else to laugh at now in the face of impending extinction by man, have at least been able to laugh at that land-based largeness limit.

I doubt if shrews have skeletal worries, certainly none connected with size and weight. But they do have a major worry, a metabolic one that derives directly from the combination of their smallness and the fact that, like the rest of us mammals, they are homeotherms, warm-blooded.

Cold-blooded creatures like reptiles and amphibians, unable to generate enough internal heat for an active life by their own metabolism, are dependent upon help from their surroundings. They are borrowers of warmth. When they can borrow enough, they are chipper and active; when the supply dwindles, they become sluggish edging toward inaction. Being so dependent upon their environment, unable to control their own body temperatures, they are at its mercy. If it lets them become too cold, they sink into a seemingly lifeless torpor. If it becomes too hot, they die. Their lives are complicated not only by the common pursuit of all living things, the search for food, but also by, for them, the equally necessary search for properly temperatured surroundings through the endless cycle of day and night.

As I see it, our early homeothermic ancestors made the decision in their time (and thus for us their descendants in our time) to be as independent as possible of their surroundings in regard to blood-body temperature — to be able to determine for themselves when they would be active. They did it by stepping up their metabolism to the point at which they could maintain an adequately high and a relatively constant internal temperature. Nowadays for most of us mammals that means body temperatures ranging by species between 95° and 100.4° Fahrenheit.

Maintenance is not exactly easy. We are always losing heat to or gaining heat from our surroundings as their temperature drops below or climbs above our body temperature in the cycles of day and night and of the seasons and with the vagaries of our home heating and cooling devices. So we have what our ancestors evolved for us, built-in automatic thermostatic controls. If we begin to cool down, our bodies increase their metabolism to produce more heat — and if that is not enough, they try to add more by the muscular action of involuntary shivering. To which we can add even more by voluntary exercise. If we start to become too warm, our metabolism slows down and our sweat gland systems turn on their evaporative cooling effect.

Such controls would be fully adequate in cooperative climates, those that never varied too much up or down from our internal norms. But from the beginning we mammals have obviously wanted to live almost anywhere, to populate the whole of the world. So we have been busy through the ages evolving additional thermostatic aids. Insulations, for example: thick skins, hair or fur coverings, layers of subcutaneous fat or blubber. With us humans, clothing.

Size has an immediate application to the problem. The way in which we lose heat to or gain it from our surroundings is through our body surfaces — which includes the interior of our lungs. As a result the ratio between body volume (the heat-producing mechanism) and surface area (the heat-loss or gain medium) becomes very important.

Think of it this way: A 1-inch cube has a volume of 1 cubic inch and a surface area of 6 square inches for a ratio of 1 to 6. Double the size dimension and the volume becomes 8 cubic inches, the surface area 24 square inches for a ratio of 1 to 3. Double again and the volume is 64 cubic inches, the surface area 96 square inches for a ratio of 1 to 1½. That is: the volume rises much more rapidly than does the surface area.

For small mammals, those with volume small in proportion to surface area, the problem is primarily one of heat loss, of maintaining the proper body temperature against the constant heat loss (except in hot weather) through the proportionately large surface area. The smaller the animal, the larger the problem. For large animals, those with surface area small in proportion to body volume, the problem becomes one of heat gain, of getting rid of excess heat through the relatively small surface area. The larger the animal, the larger the problem.

A corollary of the above is that, in general, warm climates favor small animals, cold climates large animals. Which is one reason, perhaps *the* reason, why even within a single species those in the southern part of its range (if it has a large enough range) are usually smaller than those in the northern part.

Shrews have the heat-loss problem of smallness. Being the smallest of all mammals, they have the largest such problem. They very likely could not be much smaller and survive. Most of them must be close to, the pygmy species right up against, the land mammal smallness limitation.

What sets that limitation? The necessity for fuel. For food. For the fueling of the metabolic activity required to produce enough heat for the maintenance of the proper body temperature despite the loss through a surface area so large in proportion to the tiny body-engine inside it. The average shrew must find and eat and digest and convert into energy at least its own weight in food every day, the smaller species much more than that. A shrew can starve to death in a matter of hours. Time and again a naturalist has caught one, popped it into some kind of container, and continued his rambles to arrive home a few hours later to find his prize dying or already dead. Not from shock or fright. From simple lack of fuel for its internal fire.

Fire is, I think, the right word. Life comes closer to burning in the shrews than in any other mammals. Their metabolic rate is high, their pulse and breathing many times more rapid than those of, say, us lethargic humans. They live at a high pitch of intensity. They can rarely take things easy. They can rest only in brief snatches. They are short-lived in our terms, perhaps eighteen months or a trifle more at maximum. But who can say with certainty they do not really *live* as much or as fully in their short span as we do in our long-drawn three score and ten?

Thereby hangs that ferocity charge. A shrew is almost constantly active, dashing about in seemingly erratic spurts, snouting through leaf mold and ground debris and exploring the burrows of other animals, in search of, in pursuit of, fuel for that internal fire, ready to seize whatever might qualify, on occasion to attack and conquer other creatures even larger than himself. Even when standing still, which happens rarely, his flexible snout is working as if scenting for prey, his tiny ears seeming to strain to catch the slightest sound. But he is not ferocious in this. No more so than I have been when I have chopped off the head of a chicken on its way to the family pot. Simply a job to be done — and for a shrew to be re-done and re-done and re-done. His small size keeps him on a treadmill, an almost constant cycle of searching for food to fuel the search for more food to fuel the search for more food just to stay alive.

The other day in a current animal book I came on references to the shrew as a "terrifying" little animal, one that "lives by murder." If the word is to be used in that incorrect sense, what flesh-eating animal does not live by murder? Only the carrion eaters. To a considerable extent we humans live by such murder. Most of us let others do the murdering for us, then we dine with relish on the results.

Shrews happen to be, within their smallness range and under the drive of necessity, quite efficient killers. That is to say, they are efficient in the business of living. That should be cause for admiration, not condemnation.

The female is deadlier than the male. Think of the task she faces two sometimes three times a year, that of keeping her own internal fire going and at the same time those of four to ten fast-growing young who are dependent upon her until they are about half grown.

You might think the male the deadlier when mating — he bites her vigorously while mounting her. He is merely doing what apparently is necessary to stimulate ovulation in her. If he is too gentle, the mating very likely will be sterile.

Both of them might seem to be handicapped in pursuit of food by their congenitally poor eyesight. But so much of their hunting is done in the dimness of accumulated groundcover or in the dark of burrows that good vision is not very important to them. Their long widespread vibrassae (whiskers) are more useful, being wonderfully sensitive, serving as pathfinders and prey locators. Even more useful may be the high-pitched squeaks that they emit as they dash about — and that have helped tag them with the false reputation of being ill-tempered scolds and greedy murderers.

What has recently been suspected and now is fairly well established for some species (in time perhaps for more, even all) is that shrews, like bats, have evolved a sonar system. They catch the echoes of their squeaks bouncing back and decipher these into knowledge of their surroundings sufficient for aid in getting about in dim or dark places and in locating moving objects. The squeaks coming from shrews as they pursue their prey, then, are not the ferocious sounds of killers lusting in anticipation of the thrills of the kills. They are sounds helping them in their shopping for food in the super-markets of their home territories. There is probably less emotion and indication of character in those squeaks than in the squeals of housewives who have spotted special items on a bargain counter.

When not driven by the imperative need for food, a shrew is a pleasant and cheerful fellow citizen of this world. He is neat and well groomed, finding time despite always being in a hurry to clean his face with his hind feet and to wash his fur with his tongue and comb it with his toes. If kept in a cage, he will daintily use only one corner for a toilet. People who have kept shrews thus (and well fed) report that they seem to enjoy inventing small games. Konrad Lorenz, who set a record by keeping a crew of eight littermates alive and healthy for a period of months, insists they were "playful" and "good-natured."

Both of them, he and she, enjoy each other's company, not just during the mating period but before and after, on through the gestation time. It is only when the young are about to be born that she tells him to be off on his business while she tends to hers. For the next few weeks she will be a dedicated parent, fueling herself and her litter. As soon as they can get about, the young ones will follow her as if on strings, at the slightest alarm ready to form in a split second what is called a shrew "caravan." One will grab hold of her fur near the base of her tail, another the fur of that one in the same manner, and so on down the line. Sometimes, in their hurry, they form two lines. Either way they will then "play" follow-the-leader in perfect synchronization. As she leads off, they follow, staying in step. If she stops, they stop. So complete is their faith in her that she can be picked up and they will be hanging in that line (or lines) from her. When they are about half grown she sends them off on their own. She has not had to teach them much about hunting. They were born with that drive for food and the right instincts for getting it.

The claim is often made that shrews are cannibalistic. I suspect that one of them, coming on the carcass of another, might make a meal. But the only records of one shrew killing another I have yet encountered refer to definitely unnatural conditions — two or more of them cooped in even for them very confining quarters. I have come across accounts of one human killing another under similar conditions. What really happens when two shrews, say two adult males, tangle in the wild over such a vitally important matter as territory? These supposedly violent-tempered cannibalistic murderers do not even engage in a true battle. They touch vibrassae, estimating each other. Then they squeak. And squeak. They will rear up on their tiny haunches and squeak even more intensely. They will throw themselves onto their backs — and SQUEAK! At last one or the other will decide he has lost out in the name-calling or simply the hollering contest and will retire in defeat. Or perhaps the loser will have entered the lists with a less well-filled gut and has suddenly realized he is wasting precious time and must get back to the never-to-be-neglected business of finding food.

What food? Anything in the overall meat line small enough to be caught and conquered. Insects and larvae, slugs and centipedes and spiders, snails and other mollusks, earthworms and small salamanders and lizards, young mice and even their parents, an occasional small frog or toad or snake. Carrion too — and if maggots are working on it, all the better.

But always and primarily the first items in that listing: insects and their larvae. Examinations of stomach contents usually show that some 80 percent is insect matter, most of it so finely chewed that preferences in insect species can not be determined. Probably there are none — except that the larger the insect, the better the snack.

I suspect that to shrews we humans are nothing more than hazy large moving objects who sometimes kill one of them under tangled grasses or fallen leaves by stepping on him without even knowing it. To most of us they either do not exist or do so only as peculiar little beasts heard or read about. And yet: there they are, millions upon millions of them around the world, invaluable allies in our accelerating war against our insect competitors and in preventing too many population explosions in micedom. Each shrew is busily eating at least his own weight in what we regard as pests every day and in what are called "favorable locations" there can be as many such busy eaters as one hundred to an acre.

They give us aid. We give them none — and malign them in opinions of them and in definitions in our dictionaries.

Back somewhere I wrote that smallness imposes a penalty. No, the shrews say to me. No. Not a penalty. A challenge.

There they were some scores of millions of years ago, the first true placental mammals. Warm-blooded. Tiny creatures up against that relentless heat-loss problem. There were ways to ease it.

One was to become bigger. Some of them, fortunately for us, did that and in time founded the other placental mammalian orders and families. But some went right on being shrews, accepting that challenge, meeting it, defeating it. They could do it, so they did. They are still doing it.

Another way was to seek and stay in cooperative warm climates, the tropical regions. Some of them did. But more of them spread out through the temperate regions where the seasons complicate the problem and some on into the Arctic itself, where the problem would seem to be almost insoluble. They were too small to be able to hibernate. They could not fly like birds to warmer regions when the bitter cold came. They had to rely on the constant fueling of their internal fires, all day and every day the year around. They could do it, so they did. They are still doing it.

To me that speaks of a bravely adventurous attitude toward life, a loving of life for the simple fact of itself alone, a willingness to live intensely on the knife edge of existence. That attitude has kept them going while other whole families of mammals have evolved and had their times and have dwindled away — has kept them going longer than any of all the rest of us placental mammals have yet attained or may ever attain.

I look down from the eminence of my much larger size at those tiny flickers of life-flame that are the shrews. Considering and contemplating of them, I salute them for what they have achieved and daily achieve. They seem to be saying something to me, suggesting that adversity has its uses and compensations, that survival and success in living may not be aided, may even be hampered, by the pursuit of ease and comfort and security.

Order: *Lagomorpha*
Families: *Ochotonidae*
Leporidae

The Hare-shaped

EXPERTS in the field of animate evolution sometimes suggest that the whole long continuing story of life on this earth could be told in terms of a consistent trend toward an increase in the amount of living matter. There is, for example, the apparent tendency of all forms of life to multiply toward the carrying capacity of their environments. Equally important or more so, there seems to have been a fairly constant expansion of new forms of life to make use of all possibilities existing at any time and such expansion has itself in turn rather steadily opened up further possibilities.

Consider the cumulative effect of the climb of plant life out of the waters of the earth onto the land. Certainly an expansion to take advantage of possibilities. In turn and in time this created possibilities for animal life to become terrestrial too. With development of land forms of animal life came possibilities for development of predators to prey on them. Along the way for both plant and animal life the arrival of new forms offered new possibilities for bacteria, protozoa, parasites and such to do their kind of preying. And so, perhaps, as Jonathan Swift rhymed of fleas, *ad infinitum.*

Paleontologist George Gaylord Simpson puts the overall point plainly. In his *The Meaning of Evolution* he states that this trend toward a net increase in the sum of living matter, this "increase in the total metabolic processes," does appear to be "the most nearly universal phenomenon of evolution."

A corollary that could slip into an unwary mind is that we humans with our population explosion are simply doing our duty, seeking to obey an evolutionary imperative. But my mind, though often unwary, scents the fallacy there in regard to us resource-monopolizing humans and is currently content to consider and contemplate of certain small mammals who seem to me to have tried and still try to obey that imperative in a reasonable manner and with remarkable consistency and against great odds.

They are the lagomorphs, order Lagomorpha, which to Greek scholars identifies them as the hare-shaped: the hares and the rabbits and the pikas.

They have not been particularly inventive in their efforts to boost the bulk of living matter in the world. They have not radiated out into many life forms, into many families and genera. Instead they have clung primarily to the simple direct method they developed long ago: multiplication of individuals. As Goldsmith noted in his *Animated Nature,* having been provided with "more ample powers than most others for the business of propagation," they "found a resource in amazing fertility." Having found it, they have kept on doing their duty in their way with unusual generic constancy. The lagomorphs have made fewer efforts at radiative adaptations and have had fewer now-extinct life forms than any other mammalian order. The original forms, once established, were admirably adapted for widespread survival with little need for further inventiveness.

Few other animals have been as supposedly well known or as lengthily discussed from the time of Aristotle onward, yet it was only recently that the lagomorphs won recognition as comprising a taxonomic order of their own. They were formerly regarded as rodents, still so classified into this century, primarily because they do bear the rodential insignia: two pairs, upper and lower, of prominent chisellike incisor teeth that continue growing throughout life. But the lagomorphs go the rodents one pair more, have another pair of smaller incisors just behind the upper outward-showing pair.

In fact they have still another upper pair at birth, but these soon disappear. As far as I have been able to determine, no one yet has solved the why and wherefore of this adult double dentition, but teeth are too important in taxonomy for such a peculiarity to be ignored. Confronted by it, the classifiers tucked them into a suborder labeled Duplicidentata — which meant, of course, that all the multitudinous others in the then Rodentia order had to have another suborder labeled Simplicidentata.

Then more differences were emphasized. The lagomorphs have more premolars than do the rodents. Their cheek teeth are designed more for cutting than for crushing, as are those of the rodents. Their ears are always longer than their tails, the reverse of what is usually and emphatically true of the rodents. Female lagomorphs (this is what impressed Goldsmith) outdo other female mammals by being ready to, wanting to, demanding to, mate again almost immediately after giving birth. And yet another peculiarity setting lagomorphs apart not only from the rodents but from most other placental mammals is that the males, like human males, have no penis bone and the scrotum, as in marsupials, is located in front of the penis.

Nowadays it has been established that the lagomorphs and the rodents were already quite distinct some sixty million years ago. Those misleading lagomorphic incisors are not even remotely rodent-related, are the result of convergent evolution; that is, a similar but independent adaptation.

Sometimes I wonder whether prominent chisellike ever-growing incisors are not the supreme achievement of mammalian evolution. Those two orders that share them, the rodents and the lagomorphs, have always been among the most successful in terms of population and distribution and survival. Only the bats and the shrews (who both have quite good incisors too) really compete in such terms. We humans, with brain and central nervous system our boast, like to think we are the most successful of mammals and in many respects right now we are. But we have been around only a few ticks of the evolutionary clock and truly successful for much less time and only recently have we become worldwide in distribution. The lagomorphs and the rodents, special-incisor-equipped, have been extraordinarily successful citizens of the world for fifty to sixty million years.

There is always the possibility that in another fifty to sixty million years the lagomorphs and the rodents will be gone and we humans will still be here, still boasting of our brains. Eventually we might even outrun their record. Since I do not expect to be here to assess the verdict, I report my guess: when we humans have eliminated ourselves or been eliminated by whatever fate overtakes us, some of the rodents and lagomorphs will still be around, incising continued survival.

The rodent success has been achieved primarily by wide radiation into a vast variety of life forms. I suggest that the lagomorphs rate particular respect because they have achieved rather comparable success despite what could seem a handicap, a limitation to the ridiculously tiny total of just three life forms.

Lagomorpha — "hare-shaped." Of course hares are hare-shaped. The other two, rabbits and pikas, have the same general shape. The same, that is, in basic contour of head and body. The three have done their individualized tailoring with their appendages, with hind legs and ears and tails. They have followed an orderly sequence in a lengthening of those appendages. The bigger the body, the longer the hind legs and the ears and the tails in relation to that body.

Pikas are the smallest, start of the series. They have short hind legs only a trifle longer than the fore, small rounded ears, and tails so short they seem almost nonexistent. Rabbits, adding size, have hind legs much longer than forelegs, ears really reaching for length, and tails long enough to be noticeable. Hares, largest of the three, have the longest proportionate hind legs, the longest proportionate ears, and tails almost long enough to really be called tails.

The progression can even be followed within the category of rabbits, the various species showing the same proportionate lengthenings as body sizes increase. The whole sequence from little pika to largest hare is directly tied to habits and habitats, neat example of an order of animals fitting themselves for survival in widely varied environments. Such efficient fitting that long before we humans made our arrival they had spread their natural range through all the lands of the world except southernmost South America, Antarctica, the Australian–New Zealand region, and various oceanic islands. I believe it is obvious that ocean barriers alone caused those exceptions. South America provides evidence to prove the point. As long as that continent was isolated, the lagomorphs could not colonize it. As soon as the Central American land bridge became available, they made use of it and spread southward. By modern times they had populated two thirds of the continent, an expanding process probably still under way.

The final proof is that since our human arrival we have introduced some of them almost everywhere on the above exception list except Antarctica — and they have thrived so well that usually we have regretted doing so.

Contemplating of the separate lagomorph life forms logically begins with the pika, not only as the start of the structural sequence but as the most conservative of the three. He has remained, I believe, the closest to the original lagomorph ancestry. He sets the family pattern in regard to such items as the following:

That he has proportionately large protruding eyes set close to a high arched nose enabling him to see both sides and forward at the same time except for a blind spot directly ahead — for which he can make correction with slight movements of his head.

That he has five digits on his forefeet, one of these much reduced and not of much use, and four digits on his hindfeet. In effect, if he were hoofed, in this respect he would be an artiodactyl.

That the bones of his forelegs cannot be turned inward. Thus he cannot use his forelimbs like arms-plus-hands in feeding and depends for this primarily on his very pliable lips.

That his hindfeet are quite long as feet go and add to the disproportionately lengthy appearance of his hindlegs when he is up on tiptoe — one of the family's aids for leverage and bounce in running.

That his nose is almost always trembling and twitching as if he had some sort of nervous affliction.

That he never hibernates.

That his females average somewhat larger than his males.

For himself he looks like a plump small half-grown wild rabbit who has had ears and hind legs and tail drastically trimmed. In coloration he can be quite varied, running through grayish buff and cinnamon to tawny and, as usual among mammals, he is darker above and lighter below. The tips of his back hairs are often blackish and as they wear away help give him a wide variety of color tones. He renews his fur in his own way, shedding of the old and growth of the new beginning with his head and progressing backward. As to motor activity he seems to have just two gears: very active or completely at rest.

There must be a streak of independence in him. Way back in the Oligocene, when the lagomorph line was doing its relatively scant diversifying, he must have decided not to go along with his diverging relatives, rabbit and hare, in efforts to increase the amount of living matter in the world by sheer production of progeny. He had the same basic equipment for it, but he adopted a different method that, while permitting him only a brief breeding season, would enable him to show prowess in another direction. He picked an independent place in what Darwin called the "polity of nature," chose what nowadays is known as a special ecological niche. He began to populate unpopular habitats, to occupy areas where few other mammals try to survive — even some where no others make the attempt.

The response might be that he was really lazy or cowardly — that in picking unpopular habitats he was trying to avoid competition. True enough, the environments of his choice give him few competitors for food and space and reduce the array of predators with whom he has to deal. But they pit him against more pitiless competition, that of inclement weather. He has tuned himself to low habitat temperatures. He lives where winters are rough and rigorous, even where deadly winters are almost year-round. Much of his range is in regions where winter thermometer readings — if thermometers were there and people to read them — would average well below zero for months on end. He, a warm-blooded mammal with a high internal temperature to maintain, actually prefers that kind of climate. He will doze contentedly out in the open in record subzero readings. In warmer weather, what we other mammals would regard as merely not-quite-so-cold weather, he may retire to some shadowed more-comfortably-cold hide-out. He can, and some of him do, enjoy life far up on Mount Everest at an altitude of 17,500 feet, highest habitat of any mammal.

His major development has been in Asia and his major range nowadays is all the way across the northern portion of that huge continent. During some long ago time, obviously when a land crossing of what is now the Bering Strait was available, he made his way into North America. Here too he could find properly chilled regions. No doubt his range on this continent fluctuated widely through the relatively recent ice ages, expanding when the glaciers creeping southward created more areas that would please him, contracting when the glaciers retreated northward again. Today his range here covers much of Alaska, as much or more of western Canada, and scattered areas down through northwestern United States.

In the northernmost parts of his overall range, since they are adequately refrigerated, he can be quite happy all the way down to sea level. In the more southern parts he has to be selective, find his cool where he can — which, of course, is usually in mountain areas. Thus in the United States, while some of him live on high plains and in old lava beds in the northwest, most of him occupy the upper reaches of the Sierra Nevada and the Rocky Mountains, particularly the latter.

I think he likes the Rockies not just because they provide some cold but also because they are what their name states, rocky. He regards rocky regions as prime real estate. Talus slopes and rock slides are his favorite homesites. The more cracks and crannies and passageways the better. Which is why he has never bothered to lengthen his ears and hindlegs to match the other lagomorphs. He does not have their need for extra-acute hearing and high-speed running. When he is startled or when danger threatens, all he has to do is pop into a nearby crevice.

How does he do it, carry out his program of populating regions seeming so inhospitable? He is a vegetarian, yet he lives where during much of the year little or no vegetation is available to him. He disdains hibernation — is too small anyway to be able to hibernate successfully through the kind of winters he favors. How does he get through the long cold subzero months?

To begin, he is built and equipped for cold climates. His body is chunky and round with short appendages, which means that the ratio between his small bulk and his surface area is favorable, holds heat loss to a minimum for his size scale. Then too he has developed the lagomorph three-ply style of fur close to perfection: fine short very dense underfur; longer coarser hairs whose tips overlie and conceal the underfur; still longer coarser hairs whose tips overlie and conceal the middle layer. In effect he wears three fur coats at once. His tiny ears are furred on both sides, inside as well as outside. And he has nicely thick fur on the bottoms of his feet, which serves as warming moccasins and gives him excellent traction on the slickest of rocks and slipperiest of ice.

But his major resource is tucked away inside his small hare-shaped head. A knowledge. A skill. He follows an old English proverb — and was doing so long before there were any Englishmen to provide proverbs. Quite literally he "makes hay while the sun shines."

In his northernmost habitats through whatever of a greenery growing season they afford and in the more southern when he knows that winter is approaching he is at work cutting and curing his hay. Since the sun is his assistant, he is chiefly a diurnal not a nocturnal animal. He is up and out soon after dawn, scurrying about snipping with those incisors at grasses, sedges, weeds, flowering plants, whatever is available, collecting his cuttings into a bundle almost as big as himself held crosswise in his mouth. He carries his bundles to a sunlit flat surface, preferably rock, and spreads them out in little piles. Since he makes many trips, he can have quite a collection on a good day. If shadow creeps over as the sun shifts, he may move his piles to keep pace. If dissatisfied with one location, he may move them to another. If a storm approaches, he hurries to carry them to shelter. If this happens to be a night storm, he may work a night shift to salvage them. When they have been properly cured, they will go into one of his sheltered haymows. He is as knowledgeable in what he is doing as any good farmer anywhere.

Let winter snow and winter cold come. He has a seasonally renewed three-ply wrapping around a metabolically efficient little body. His barns are stocked with hay — and hay that will provide dietary variety because it is composed of a variety of plants. Though heavy snows and slides pile a blanket ten, twenty, forty feet deep over his homesite, he is content and well-fed below in the rock hollows. Out of adversity he has created comfort. No wonder he likes real winters. They are his leisure times. If he wants a bit of diversion, he can tunnel out through the snow to some wind-swept bare rock. He can recline there at ease, looking out over his chosen frozen realm in simple assurance that he has earned his refrigerated rest.

He is alone there on his rock and you might think he is lonesome. He is not. He is part of a neighborhood. There are others on other rocks not far away. He is both territorial and social, and has worked out a sensible combination. He insists upon the privacy of his own curing ground and haymows and home, but he likes to have companions with adjacent homesteads. Since they are small citizens, their personal territories are small, relatively close together, and they regard all other surrounding areas and forage grounds as completely communal. I am reminded of us humans who call our homes our castles yet like neighbors nearby and are quite willing to jostle each other in the workaday world.

He and his neighbors have, too, a communal communication system. The vocabulary is limited but well used, the major sounds a mutual warning signal, a high-pitched sharp bark or whistle, and a now-all-clear reassurance, a series of short bleating calls that dwindle away.

He must put real effort into such signals — at least he seems to, because his little body jerks forward and upward with each call as if he is forcing it out. Perhaps he is concentrating on a trick he has acquired; somehow he manages to make his voice ventriloquial. He can be announcing your approach to his companions, loud and clear, yet you will look for him everywhere he is not before you may manage to catch a glimpse of him. If he has popped down among his rocks, his repeated sharp "ka-ack" will seem to be coming from a dozen places at once.

No doubt he long had a variety of vernacular names across Asia, just as he did in the languages of some western Amerindian tribes. But in these modern supposedly more civilized times he was late in acquiring any name at all — and when he did, he had scientific labels before he had a decently acceptable common name. Let me see if I can straighten out the record enough to make some sense.

What might be called the first official notice of his very existence came in 1769 when a description of a Siberian species was published and a Latin tag supplied meaning "tiny rabbit." He had to wait until 1828 for first notice of an American species. Meanwhile knowledge of his Asiatic representatives had become sufficient for his right to a separate genus to be recognized and *Ochotona,* based on a Mongolian name for him, was suggested. The suggestion did not take and for quite a while *Lagomys* from Greek roots meaning "rabbit-mouse" (or "hare-mouse") was used. Eventually someone pointed out that *Lagomys* had earlier been applied to the marmot and therefore was pre-empted. So the taxonomists returned to the Mongolian with a new spelling, *Ogotona.* Finally and at last they agreed on the original *Ochotona.* And then, recognizing his right to a separate family as well as genus, they built on that genus name for the family name, Ochotonidae.

There remained the problem of differentiating the Americans from the Asians. Along the way a name for him, "peeka," used by the Tungus tribe in Siberia, had been offered for the genus. The now prevalent priority rule blocked such use but there could be another use. When the dust of disscussion settled, *Ochotona* was definitely the genus, "peeka" become *Pika* was a subgenus for the American species.

While the taxonomists were playing their labeling game, ordinary folk were becoming acquainted with him by meeting him in our western states or reading about him in travel books. Since it is impossible to have a good look at him without seeing the relationship to the rabbit-hare complex, he was usually regarded as simply a small edition. Thus the names variously given him: rock rabbit, rock cony, squeak rabbit, calling hare, mouse hare, piping hare, rat hare, whistling hare. Some people did a bit better by referring to his occupational specialty with haymaker or rock farmer. But to keep on using any of all such would be an insult to his individuality.

For some time all his true friends, including most naturalists, have used the subgenus label as his common name. With no geographical limitation. Whether high in the Himalayas or atop an American peak, he is a pika.

The biological conservatism of the pika seems to me emphasized by the fact that he, a quite ancient very small mammal with far-scattered often climatically isolated habitats within a vast range, circumstances usually encouraging genetic inventiveness, has held his family to just the one genus and only fourteen species.

Twelve of the fourteen are found in Asia, two in America. Of these *Ochotona collaris* inhabits Alaska and northwestern Canada. As his name indicates, he wears a collar, a darker patch on the back of his neck and shoulders. The other American species, *O. princeps,* lives in southwestern Canada and down through the western mountains of the United States.

That species name, *princeps,* was derived in a roundabout way, translated first from the Chipewyan Indian into English as "little chief hare," then from that into the Latin for "chief." This one has a slightly narrower and longer head than *O. collaris* and is usually gayer and brighter in coloration.

Because of his preference for regions rarely visited by us humans the pika is not well known and the general impression may be that there are not many of him. Actually, wherever he does live he is usually well represented. He is, after all, a lagomorph. Though the brevity of his possible breeding season limits his mate to one brood a year, she averages at least four young and takes good care of them. They are born naked and helpless, but within six to seven weeks are almost full grown and can expect a life span of up to three years. The combination of a relatively limited list of predators and a relative ease of escape from them helps assure a good survival rate. "My own travels," wrote Ernest Thompson Seton, "have made me familiar with the high mountains from central Colorado to central British Columbia; and within this, I have never yet seen slide-rock up near timber-line without finding the little bump that squeak-squeak-squeaks from some low point of rock."

Trying to photograph some of him one Sunday morning well up Pecos Baldy here in New Mexico, biologist Bailey learned too that the vocal little bump is not overawed by a looming human presence. After the initial alarum at the intrusion, the various haymakers returned to their haying. One "old fellow" kept right on, though Bailey had taken a camera position close to his current hay piles. Once he took time out to approach within two feet to inspect the intruder. Occasionally he asked apparent questions of intent and was reassured by soft answers. At last, coming back from the forage ground with a mouthload too big to see around, he ran headlong into one of Bailey's boots. Obviously he knew his terrain so well that he could scamper over it blindfolded and to run into something new suddenly there was something shocking. He dropped his load and dove into a crevice. For a few moments his indignant remarks came from deep in the rock pile, then he emerged to continue his scolding from a nearby vantage point before going back to work.

That was early in August. Later in the year with winter snows threatening, the same "old fellow" might have been both more unafraid and more indignant. Late season researchers checking haypiles to see what plants they contain have been surprised to have the owners come running to scurry around the gross human feet, trying frantically to restore the piles while these were still being pulled apart for inspection.

Snow is falling outside the window of the room in which I write this. It does not bother me. The house is weather tight with a competent heating system and there is ample food in kitchen cupboards. I can claim no credit. I did not build the house or install the heating system or harvest the food. I am dependent on others for those things. I am even dependent on a power company to continue supplying the current which enables that heating system to work.

On out the window through the swirling flakes I can catch glimpses of the Sandia Mountains, their tops white-mantled and cloud-hung, seeming to frown an ominous warning of worse to come. I am glad I am not up there right now. And my mind lingers on a small stubborn self-sufficient squeaking bump who, if he did any thinking on the subject, would also be glad he is not up there — for precisely the opposite reason. Though the Sandia Crest tops out at more than ten thousand feet, it is not cold enough long enough up there for him. The snow does not pile deliciously deep enough. Even in the most hidden canyons the snow is usually all melted by June instead of staying around into August and perhaps until the new snow of the new winter starts to fall, as it does in proper pika country.

I cannot see them from here, but I know that sixty-some miles to the north the Sangre de Cristos start their tremendous march northward into Colorado. I know that from the triple Truchas Peaks behind Santa Fe on up to the topmost of them all, Wheeler Peak above Taos, wherever they rise through the Hudsonian on up above timberline into the Arctic-Alpine, there he is, little chief *O. princeps,* snug among his comfortably cold rocks, nibbling his variegated hay. The credit is all his.

The other lagomorphs, the rabbit and the hare, resemble each other so closely that they complete the order with just one family shared between them, Leporidae, and are usually discussed together as the leporids. In taxonomy they are separated at the genus not the family level. Even so they are separate enough, I insist, for me to refer to them as separate life forms.

Experts can cite many intricate structural differences between them beyond that of proportionate length of appendages, particularly in regard to skull bones. But the major easily noted difference is that the rabbits make burrows or other nests and produce young blind and naked and helpless in need of much maternal care while the hares do not bother with nests and use a longer gestation period to produce young eyes-open and furred and able to run about within a few minutes and to take care of themselves in quite a brief time.

Since the leporids preceded us humans and were known to us from the beginning of our existence on earth, there was ample time for supposed knowledge of them and names for them to be wonderfully well mixed up before modern scientists began to make sense of it all.

The classical Greeks used the one term *lagos* for both of them, apparently regarding them as varieties of the one animal. I doubt whether Aesop, when writing the four of his fables featuring leporids, knew or, if he knew, thought it made any difference whether they were rabbits or hares. The Romans must have known there were two of them because they had two terms: *cuniculus* for the rabbit, *lepus* for the hare. Knowing there is a distinction is not the same as recognizing it in regard to individual animals. I am fairly sure the Romans were as confused in application of the two terms as most people still are nowadays.

Our versions go back to the Old English *hara* and the Middle English *rabet*. Popular usage of the two names has always been indiscriminate, treating them as if they were virtual synonyms. And so most of us go right on referring to some rabbits as hares and some hares as rabbits. The Belgian hare, for instance, is really a rabbit, while all jack rabbits and the snowshoe rabbit are really hares.

Adding its bit to the confusion is that name cony, which I cited in my list of onetime names for the pika when he was regarded as a rabbit. It was a popular Elizabethan word for rabbit (Shakespeare used it) and is still rather extensively used, particularly in England. But the true cony is another small animal (also known as the hyrax) who inhabits the Near East. Though he has some external resemblance to both the lagomorphs and the rodents, he is actually so different that he has his own tiny taxonomic order.

I suspect that what helped clinch cony into English usage was the King James Version of the Old Testament. In Deuteronomy 14 a passage listing animals "ye shall not eat" was correctly translated as "the camel, and the hare, and the coney." I do not know what was in the minds of the translators when they put down "and the coney," but I am reasonably sure that most readers of the time, who had never heard of the true cony, regarded the phrase as the equivalent of "and the rabbit." On the other hand I am absolutely certain that the original Hebrew writers meant the true cony, an animal with which they were definitely familiar.

Obviously they were also familiar with a resource of the hare (and perhaps the cony) of which I, though a onetime hare-raiser, had no notion until I became curious about that Biblical passage. Why should ye not eat of the camel and the hare and the coney? Because "they chew the cud, but divide not the hoof."

Of course the hare has no hoof to divide. But does he chew a cud? Yes, I know now — in his own way. Most leporids do when food is scarce or that available is hard to digest. They practice what is called coprophagy, the re-ingestion of fecal matter. Food passes rather rapidly through their digestive systems and, after the first passage, is voided in soft greenish pellets, which they may promptly re-eat for a second chance at extraction of its nutrients. What remains after the second passage is voided in brown hardish pellets, familiar to all leporid fanciers. In a functional sense that does resemble the cud custom of the camel and other ruminants. The hare, who frequently lives in sparsely vegetated semi-arid regions, rather regularly practices it.

I do not know whether the true cony does the same. But I have turned up statements that he works his jaws "in a manner reminiscent of cud-chewing."

The confusion of names is as nothing to the confusion of notions that early accumulated then persisted with additions through the centuries. I am content to mention only a few taken from Topsell's *History of Four-footed Beasts,* which always supplies fascinating reading. Being a late Elizabethan, he naturally had a flavorous writing style and also called the rabbit a cony — and considered him one "among the divers kinds of Hares."

Efforts to explain the leporid ample powers for the business of propagation gave birth to a series of notions. Topsell traced the start to the Hebrew word for hare, which, being in the feminine gender, "gave an occasion to an opinion that all Hares were female, or at least that the males bring forth young as well as the females." On ancient authority he cited another opinion that some hares are female only but all males are double-sexed, are also female. Yet another, that hares share the reproductive burden in alternate sequence, each individual being male one year, female the next. Not quite able to accept any of these notions, Topsell settled for another: that the female when pregnant continues "provoking the male to carnal copulation" and thus continues to conceive and always has young in various stages of development within her, giving birth to them in what nowadays could be called an assembly-line method.

More than a century and half later Topsell's version was still current with modification. Goldsmith, for instance, believed that the female could at least double the usual mammalian procedure, could conceive a second litter while a first was on its way.

Not long ago I came on a relatively recent variation on that general theme of cooperation of the sexes. I was reading a plump little volume titled *Sporting Adventures in the Far West* by John Mortimer Murphy, published in 1880. Up popped the statement that the males of "a Rocky Mountain variety" of jack rabbit have teats and help suckle the young. I will not forgive Mr. Murphy his boasts of the prodigious slaughter of such game he perpetrated in the name of sport — but I think I understand his biological mistake. Like so many others going back to Topsell's ancient authorities, he probably mistook the male testes in their scrotum for teats. Their forward position and the fact that when their owner is not sexually stimulated they are partially withdrawn into the abdomen could be misleading. At least to a gullible mind overheated like its owner's gun with the pleasure of killing.

Like the pika, leporids have hair on the pads of their feet, though not of the same thickness and quality. They also have a peculiar facial arrangement, which prompted Aristotle to state that the hare is "the only animal known to have hair inside its mouth and underneath its feet." That "inside the mouth" business is a matter of definition. Leporids do have lip folds separating the incisors from the mouth cavity when closed and these have a fine velvety covering. Let Aristotle have it, then, that there is hair inside the mouth — behind the incisors anyway. Leporids also have a Y-shaped naked (hairless) groove that runs from the upper lip to and around the nose, leading to sensory pads, which are usually hidden by folds of hairy skin, at the entrance to each nostril. The whole arrangement has prompted the expression "hare-lipped." With usual human willingness to blame anything blameworthy on others, when one of us is born with a split upper lip, some of us want to blame some innocent leporid. Topsell put this bluntly: "If a Woman with childe see one of them suddenly, it is dangerous, if the child prove not Hare-lipt." Similar notions about other birth defects or marks are as current today in some quarters as was that one in Topsell's time. As a matter of fact, I think that one still lingers.

I doubt whether others linger, such as the belief that leporids who lighten the color of their coats toward white in winter do so by eating snow or that when pursued they always run uphill because then their hindlegs longer than the fore give advantage. There was some naive plausibility in those notions. I see little, however, in the companion notion that hares were really dangerous animals because old women witches, with the aid of the devil, most frequently took their form when determined on devilment. And I see none at all in the long-held belief that leporids (being obsessed with sex) would mate with other kinds of animals. That one bred wonderful tales.

When Xerxes I was preparing to invade Greece, a mare gave birth to a hare. Xerxes should have heeded that obvious warning, but he went ahead and was defeated. Topsell saw nothing tall in that tale. He had already swallowed another and taller supposed to have happened close to his own time. In this a three-year-old mare brought forth a hare that began to speak as soon as it was born. Then it began biting its mother, sucking her blood, finally killing her. Then feathers grew out of its back and after speaking more cryptic words it flew away. One Petrus Toletuc of Lyons interpreted the message: "The days shall come except the mercy of God prevent them, that children shall think they do obedience to their Parents if they put them to death."

Topsell had to believe: "There were present at the sight . . . seven publick notaries, which called witnesses and made instruments thereupon."

Chuckle at such if you wish as tales out of the age of miraculous omens. But on into the Age of Enlightenment you will find Goldsmith, following the famous French naturalist Buffon, seriously asserting not only that "nothing is more frequent" than the mating of rabbits with hares (the result "like all other mules is marked with sterility") but also that "the rabbit couples with animals of a much more distant nature." He cited rabbits born with horns and "a creature covered with feathers and hair . . . said to be bred between a rabbit and a hen."

None such are in the same category with the famed "jackalope" of New Mexico and other western states. That wondrous creature bred of adultery between a jack rabbit and an antelope (pronghorn) does his prodigious leaping only in tales so tall it takes two people to see their tops.

The confusion of notions in turn is as nothing to the confusion and multiplicity of medicinal properties attributed to the flesh and assorted organs of the leporids. Topsell devoted type-filled pages to this and the wonder is, even if only a portion of the claims were true, that mankind should have continued to suffer physical ills. With the leporids obligingly using their powers of propagation, materials for cures of most human ailments should have been readily available. Not merely cures for ailments, but inevitably also, given the leporid reputation for sexual prowess, all manner of aids for increasing human beauty and virility and fertility.

My own theory for the failure of the leporids thus to ease human misery and promote human happiness is that most of the prescriptions were so complicated, the ingredients so repellent, and the methods of preparing them so much the same, that most people have preferred to endure their ailments and their sexual inadequacies rather than to attempt the cures.

More pleasant to my mind are the literary uses to which the leporids have loaned themselves. They were neither formidable enough nor exotic enough to be included in the original *Physiologus* or the many medieval bestiaries, but they have figured in fable and folklore all the way from Aesop's nap-taking hare through the Easter Bunny to Peter Rabbit and Uncle Wiggily and the contemporary Bugs Bunny. They have served as varied symbols from lust to innocence, from meanness to sweetness, from foolish braggart to clever escape-artist, from silly simpleton to shrewd bargainer. Though a common opinion of them is summed in the derogatory expression "hare-brained," I think on the whole a more prevalent and enduring attitude toward them has been tinged with a touch of respect and an urge of affection. For myself, I lean to what Xenophon said long ago: "No man seeth a Hare but he remembreth what he hath loved."

Here in my part of the country rabbit is a frequent character in folk tales of the Pueblo and the Apache Indians. Coyote is often after him, of course, but rabbit as often defeats that notorious trickster. I do not know whether this is the result of affection for rabbit

or of fondness for making coyote the butt of jokes. No matter, it
has prompted ingenious tales.

Among the Pueblo tales are some that surprised me when I first
came on them in publications of the American Folk-Lore Society.
Out here, far from the habitat of Uncle Remus, the saga of Brer
Rabbit and the Tar-Baby had Indian incarnations. Except that a
human farmer instead of Brer Fox concocted the sticky scheme and
the Tar-Baby was a piñon Gum-Baby, the basic framework of the
tale was identical in all its variants. Uncle Remus had Brer Rabbit
escape by tricking Brer Fox into tossing him into a brier-patch. At
the Taos Pueblo a new dimension was added: Rabbit was caught all
right and made into a soup with chili. Then, despite the warning
the master of the house gave that not a drop must be spilled, always
a drop was spilled — and Rabbit came to life and ran away.

There is a metaphoric germ of truth in that Taos tale-finish. A
tribute to rabbit — and hare. More of them have been made into
meals for a wider roster of animals and people than any other creatures above the size of the mouse, yet always they seem to have come
to life and run away to exercise their powers of propagation.

The ochotonid, the pika, made classification fairly easy by confining
himself to one genus. The leporids, the rabbit and the hare, offered
more difficulty and full agreement still evades. Checking authorities
against each other, I tally nine genera for the family, of which three
are out and away the leaders. The other six (five have just one
species each) may have had flourishing heydays in the past but nowadays are restricted in range, most of them so much so that some
taxonomists group them in a subfamily as relicts, remnants of genera that have dwindled close to eventual extinction.

Of the three leading genera *Oryctolagus*, the Old World rabbit
(from whom the domesticated breeds have been developed), also has
just one species — but with that one and with the aid of man has
more than outdone his duty. His original range was southwestern
Europe and northwestern Africa back when those regions were
much more lush and attractive to vegetarians than they are today
after millenia of human exploitation and devastation. Apparently
he spread through most of the rest of Europe on his own. Then, as

modern times approached, the Normans took him to England after
the Conquest and soon he was as English as any Englishman. Hunting him became a British sport and the hound known as the beagle
was developed for it. Hindsight has since revealed that taking him
to England was the start of a whole series of mistakes.

Warnings could have been culled from Topsell's pages: references
from the works of ancient historians to population-explosion problems with rabbits (ancestors of contemporary *O. cuniculus*) in lands
around the Mediterranean. Further warnings could have been
found in ecological changes taking place in England itself. Weasels
and stoats, dining on rabbits, were increasing in number — and as a
sideline causing more trouble for poultrymen. Pasture lands and ce-
real crops, cafeterias for rabbits, were beginning to suffer. Some
wildflowers, once common, delicatessen fare for rabbits, were becoming rare, a few extinct. Such signals, not yet severe, were ignored.

So too was the major warning which was actually so well known
that it had become a part of folklore. Everyone knew or ought to
have known that *O. cuniculus* had the habit of trying to boost the net
total of living matter in the world by multiplication of progeny and
self-sacrificial feeding of predators. The female, ready to conceive
again within an hour or two of giving birth and with a gestation
period of thirty-one days, could produce in a benign climate up to
ten litters a year averaging six young per litter. Each of these young
could be of breeding age in six months.

Theoretically one *O. cuniculus* and mate could account in three
years for about thirteen million descendants. The mortality rate, of
course, is normally very high, approaching 90 percent. Apply that
to each of the generations through the three years — and the remainder is still up in the hundreds of thousands. Whittle that down
for less benign climates, which reduce the number of litters per year,
and the possibility still stands of some thousands of rabbits from the
one starting pair.

Take two species of mammals, *O. cuniculus* and *H. sapiens*, mix well
and let them loose in the world: a simple recipe for complex trouble.

What could have been expected to happen happened. Almost everywhere the British went, the Old World rabbit followed, deliber-

ately introduced to add a touch of home or provide a source of sport. His empire expanded right along with the British Empire.

A full fat volume would be required to cover just the bald facts of the results and repercussions: rabbit population explosions scattered around the world, serious enough anywhere but particularly so on a large scale in portions of Australia and New Zealand, where cooperative climate and lack of native predators gave such explosions almost nuclear power; attempt after attempt at rabbit control with each not only failing but producing its own problems; drastic ecological changes with green lands becoming barren and varieties of native flora and fauna facing extinction; laws piled upon laws and bounties upon bounties; economic conflicts between people making money out of rabbits and people losing money because of rabbits, etc. etc. etc.

The current and continuing chapter for such a volume began in 1950 when experiments in Australia revealed that myxomatosis, a disease of certain Brazilian rabbits to which most of them had become immune, was deadly when transferred to *O. cuniculus.* Since the carriers of the virus are mosquitoes and fleas, rabbit controllers were soon busy breeding mosquitoes and fleas, infecting captive *O. cuniculi* with them, and releasing the captives into the wild with cargoes of carriers. The process was adopted in one afflicted country after another and reverse results were almost as drastic as the original population explosions. Rabbits died by the millions. And then, of course, new problems promptly popped into being: how to protect domesticated breeds of *O. cuniculus* from the ubiquitous disease carriers; what to do about predators suddenly deprived of the prey to which they had become accustomed; what compensation for people with businesses based on wild rabbits; what answer to complaints from beagle clubs whose sport was threatened; again etc. etc. etc.

Whether there will ever be a final chapter is anybody's guess. What if *O. cuniculus* develops increasing immunity to myxomatosis as his Brazilian relatives have done? He could be off on another round of explosive attempts at duty doing. He might be considering some such attempts right now here in this country where, despite the long-established Lacey Act restricting the introduction of warm-blooded animals, he has a few fairly secure footholds and we have our own beagle clubs wanting to help him have more. If he does try to explode here, I hope our more sensible and more restrained native leporids can meet and overcome the competition.

The whole unfinished chronicle is just one example of our human propensity for leaping before looking, for constantly refuting the fine scientific label we have given ourselves. To date our supposed sapience has been superbly successful at messing up the world in which we live.

It is only fair to *O. cuniculus* (who is, after all, the original Easter Bunny) to point out that we humans have been more to blame than he has been for the trouble he has caused us. We virtually invited him to become a pest — and not just by introducing him into congenial new habitats. At the same time we were busy making them even more congenial for him. He thrives best in what ecologists call impoverished areas marked by depleted and eroded soil. Such areas offer excellent sites for his intricate many-burrowed warrens and a good supply of varieties of vegetation that are weeds to us but favorite foods to him. By clear-cutting of forests, by consistent overgrazing of pasturelands, by poor farming practices of all kinds, we encouraged him to spread and multiply and make a nuisance of himself. In recent years, particularly in New Zealand, there have been practical demonstrations of a direct relationship: misuse the land and the rabbit population increases; improve the land and the rabbit population diminishes.

I am tempted to formulate a defense for him. His population explosions have been instinctive attempts on his part to offset our mistakes. While we, by shortsighted methods of land use, were reducing the amount of living matter it could support, he was trying to bring the total up again.

The second of the three leading leporid genera could be called collectively the New World rabbit since the genus belongs wholly to the Americas: thirteen species whose combined range extends virtually coast to coast from southern Canada on southward, deep into South America. The classifiers label the genus *Sylvilagus,* a mixture of

Latin and Greek that translates as woods or forest rabbit and is only partially accurate because the various species inhabit not only woodsy areas but every other kind of environment the Americas offer except the pika's private domain of mountain tops.

We have eight of the thirteen species in the United States. Four of these have rather restricted ranges in differing sections of the country and are known as the pygmy rabbit, the brush rabbit, the marsh rabbit, and the swamp rabbit. The other four are collectively the American rabbit of rabbits, every one of their combined forty subspecies unmistakably a cottontail. Among them the four species inhabit all of the forty-eight adjoining United States with much overlapping of ranges and survive successfully from below sea level (as in parts of Death Valley and the Salton Sink) to far up the taller mountains. *S. transitionalis,* the New England cottontail, and *S. floridanus,* the Eastern cottontail, cover virtually all of the eastern half of the country and the latter of the two stretches his name to push into some of the western states. *S. nuttallii,* who has to be content with the name of Nuttall's cottontail, and *S. audubonii,* the desert cottontail, cover virtually all of the western half of the country.

With such a wide specific array (there are three more species with a score of subspecies in Mexico) the cottontail can vary in size, but he is never really big, rarely if ever weighing more than four pounds. Even so, of course, he is larger than his lagomorph relative, the pika, and thus, following the family structural sequence, has definitely and proportionately longer appendages. He needs them, the ears to catch sound, the white under-tail for signaling, the long hind legs for speed and agility. For all his seeming shy and apprehensive manner he is quite a venturesome fellow. Although every meat eater from the weasel up to the puma and humans and including all snakes and raptor birds of any real size regards him as eminently edible, he is no huddling stay-at-homer. Though he often dives into a burrow or thicket or brier patch to evade a pursuer, he also puts reliance on running. No doubt it was his development for mobility that helped him acquire his extensive overall range.

In extent of range, in numbers, in importance from almost any perspective, the cottontail is the standard of American rabbitdom.

He is also more familiar to most of us than any other wild mammal — except perhaps for that aggravating but likable when befriended foreign importation, the house mouse.

His soft lagomorphic fur varies from reddish to greyish brown above, shading to rather dingy white below — but always pure white on the underside of his stubby tail. When flipped up that tail becomes in contrast to the brown of his body what Ernest Thompson Seton described in his tale of Raggylug as a "shining snow-white beacon." When he is not in motion, quietly resting or crouched down in hiding, his tail is down and the beacon does not show. He blends into his surroundings and is difficult to discern. But the instant he starts to move, up flips that tail. He appears to be deliberately disclosing his whereabouts. Why?

In Seton's story mother Molly's flipped-up tail is a signal guide in any emergency for young Raggylug to follow as he scampers along behind her to some safe haven. No doubt that is one of such a tail's major functions. No doubt, too, Molly and other mothers use it to lure predators away from proximity to hidden young ones. I suspect it is also a device used by all cottontails to confuse all predatory pursuers.

The cottontail is a good runner — if his bobbing hopping high-gear gait can be correctly called running. But he is not as good a runner as popular notions make him. He depends on maneuverability as much as on speed. Frequently he has to, because he is smaller than the majority of those who prey on him and not quite a match for some of them in straightaway racing. Therefore he wisely avoids straightaway races — unless he has a good lead and can afford to head straight for a known refuge. When closely pursued he is deliberately erratic in flight: dodging, weaving, bobbing, constantly shifting (or seeming to shift) the angles of his escape route. The pursuer, made overeager by and intent on that shining beacon, is constantly receiving conflicting signals from it and losing tiny fractions of seconds of pursuit correcting directions in his own forward rush. This is a case, it seems to me, of the cottontail deriving advantage from self-advertising. According to that tail he is going this way — no, that way — no, quite another. Again and again while watching a dog chase him just a jump or two behind, I have seen that dog hesitate, slow down, sometimes stop momentarily baffled, trying to figure just where and in what direction his quarry has gone now.

In their own way the cottontail and his mate are territorial animals with sexually separate rules. According to my amateur observation neither sex is much interested in territoriality during the nonbreeding season when the weather is unfavorable and food scarce. But during the breeding season, which averages up to seven months or more in this country, giving an enterprising female a chance to produce three or four or more batches of young, they both become territory-minded. He establishes a territory that may cover up to one hundred acres. Within this or overlapping into it may be the territories of two or three or more females averaging about twenty acres. He strives to maintain his area against male intruders but is more than tolerant of any females within it. Each female guards her smaller area against intruders of her own sex and is quite happy to have the male in the neighborhood because, being a leporid lagomorph, she wants to mate again almost immediately after giving birth. It is an arrangement that seems to work well for both of them.

It is also an arrangement that we American humans should heartily approve — for cottontails. Their system of breeding-season territoriality may well be what has protected us from rabbit plagues like those the Old World rabbit has inflicted in Australia and elsewhere. In terms of the Goldsmithian "amazing fertility" the cottontail is similarly capable of multiplying toward infinity. But the insistence on territories imposes a brake. The Old World rabbit is gregarious, quite social, quite willing to live in warrens that resemble crowded city slums. When raising young, the cottontail prefers a private homestead with adequate space about. That preference helps limit population and keeps it spread out, not concentrated in specific areas.

An interesting point here is that predation is not the limiting factor it was once thought to be. Under natural conditions the cottontail population of any given area remains remarkably constant regardless of the predation pressure. There is some evidence that it fluctuates somewhat in a nine-to-ten year cycle that is more marked the further north the area is. While the reasons for this cycle are not yet known, what is known is that predation is not one of them. That is: an area's carrying capacity is roughly determined within the cycle by the space available, the number of territories (chiefly female) that can be accommodated, not by predation. Or, to put that another way, it appears that predation and disease and accident tend simply to eliminate the extra population, rarely to reduce the basic population below the approximate territorial capacity. The extras, the footloose without territories and thus without established escape routes and havens, are more exposed and become much more frequent victims.

Another point that can be made, I believe, is that the Old World rabbit is the more destructive animal. He is addicted to extensive burrowing. His warrens can become positive labyrinths that encourage erosion and cave-ins. The cottontail may make an occasional small burrow with several escape hatches, but is more often content with natural nesting places and hide-outs. Moreover, since he and his are territorially spread out, his impact on the environment is usually almost negligible, merely a part of the ecological balance.

Thus, because of the combination of amazing fertility and territorial limitation, wherever we Americans leave possible habitats, we almost always have cottontails sharing citizenship with us — and only when we ourselves are to blame do these fellow citizens cause us any real trouble and even then never in plague proportions.

With none of the above am I suggesting that our national rabbit population is a small one. Spreading out territorially may impose limitations, but our 48 adjoining states provide an immense amount of space in which that spreading out takes place. Ernest Thompson Seton is the only naturalist to my knowledge who has ever attempted an overall rabbit census. For the year 1923 he came up with the astonishing figure of 500,000,000. Since I think his calculative method was faulty, I would say that figure is valid only for the possible rabbit population at the time Columbus sailed the ocean blue. Any attempt at a current calculation would have to take into account various factors added by us humans.

I do not think that as hunters we have had much effect. As such we are simply part of the general predation that apparently does not often in any area reduce the rabbit population below the natural carrying capacity. There have been increasing numbers of us and thus increasing numbers of hunters, but that increase is offset by the reduction we have made in the numbers of rival predators. The deadliest effect we have had on rabbitdom has been the constant and always accelerating destruction of rabbit habitats as the blight of our steel and concrete civilization has expanded throughout the land. And we have added a new death device in the form of the motor vehicle. Leonard Lee Rue in his book *Cottontail* supplies the data that "Missouri, the number one cottontail state, annually loses ten rabbits for every mile of road in the state." The road-kill for the country as a whole, he estimates, "may well run into the tens of millions."

Pondering such factors, I come up with the tentative conjecture that our present rabbit population just about matches our own population, and is in the neighborhood of 200,000,000.

My reference above to hunting pushes me to another point in praise of the cottontail. In number of hunters hunting him, in number of hours of so-called hunting sport he provides, in number of pounds of meat he supplies to the national diet, he is undoubtedly the most important of all our game animals.

Game management people in particular must appreciate him — and show this by virtually ignoring him for the probable simple reason that he presents them with no problem. He is definitely an important game animal. But he needs no management. He and his manage themselves. As long as we leave them some space for their territories, they will continue to do so.

The third of the leading leporid genera includes all of the hares and was labeled *Lepus* by Linnaeus long ago. The total of species assigned to it has nowadays grown to thirty. Worldwide though they are in distribution, they are all still grouped in that one genus because they have not developed much real diversity and all are relatively close relatives. I start to read down a list of their common names and note at once the international flavor: Arabian hare, Burmese hare, East Chinese hare, European hare, Indian hare. Obviously the contemporary hare life-form is so efficient that, like the rabbit, it can get along very well without much adaptation in a wide variety of environments.

As a group the American hares obey the lagomorph structural sequence. Being larger than the rabbits, they have proportionately longer appendages, particularly their ears. With the exception of a few species of bats, they have the largest ears in relation to body size of all mammals. Thereby that name jack rabbit for most of them, which was originally jackass rabbit.

They are built, with the long hindlegs the key factor, for speed, so much so that their prowess in this respect has become legendary. Mark Twain's comment about a western jack is the classic: "He dropped his ears, set up his tail, and left for San Francisco at a speed which can only be described as a flash and a vanish! Long after he was out of sight we could hear him whiz."

The hare gait (more accurately the lagomorph gait but most highly developed by the hares) is a combination of that of the horse and that of the frog. The forefeet come down separately, clop-clop, like those of a cantering horse; the hindfeet work in unison like a synchronated pair of springs. At a walk the hindfeet come down just about alongside the fore. As the pace increases they come down further and further ahead of the forefeet — or rather where the forefeet have just been. At full throttle the forward motion is a series of prodigious bounds averaging, according to species, from fifteen to more than twenty feet. In flight most hares toss in periodic "spy-hops," leaps almost straight up, presumably to scan the terrain ahead or spot the location of the pursuer.

Hares are fast. They place great reliance on running and are able to outdistance many predators — unless these are hunting in relayed pairs or packs. They can attain up to thirty to thirty-five miles per hour and some under great urgency can hit forty-five, which are highly respectable animal speeds. But speed itself is not enough (greyhounds can do better) and is no adequate defense against those deadly enemies, the big raptor birds. I believe those hindlegs have been developed as much or more for maneuverability. Even more so than do those of the rabbit, they make their owners artists at dodging, at zigzagging, at shifting directions anywhere in almost a full circle in a single leap.

In my opinion hares are not as fast as they *seem* to be. They create an impression of traveling faster than they really are. This begins at the instant of a start. They explode from crouched-down absolute immobility into full career with the first bound. No gathering of momentum, no picking up of speed. From zero to top notch in the single fraction of a second. Biologist Elliott Coues described this far better than I could.

> The first sign one has usually of a Hare which has squatted low in hopes of concealment, till its fears force it to fly, is a great bound into the air, with lengthened body and erect ears. The instant it touches the ground, it is up again, with a peculiar springy jerk, more like rebounding of an elastic ball than the result of muscular exertion. It does not come fairly down and gather itself for the next spring, but seems to hold its legs stiffly extended, to touch only the toes, and rebound by the force of impact.

It is that rebound action producing an effect of almost effortless onrush that completes the impression of tremendous speed suggested by the initial instant velocity.

In general with some variation by species hares seem to be sociable, groups of both sexes sometimes getting along well together — except in the mating seasons, when males do considerable sparring with each other to claim female favors. Neither sex seems to be territorial in the sense of driving off others, but both do have preferences for home areas. When chased they will start off at high speed (as if heading for San Francisco), but soon will be circling to stay within the known and familiar space. Both are happily promiscuous as opportunity offers and indulge in leaping and chasing courtship antics. He apparently feels he has completed his familial duty by impregnating her. She in turn takes motherhood rather lightly. Being a hare, she can. No nonsense about nests and other preparations. She gives easy birth wherever she happens to be at the time, suckles the young ones and makes some effort at hiding them, and returns thereafter at night to feed them. Within about two weeks the young can (and often have to) survive on their own.

L. americanus, the snowshoe rabbit sometimes and more accurately called the varying hare, is the smallest of the group, yet he outdoes all the others in extent of range, inhabiting most of Alaska, most of Canada, the northern United States from coast to coast and on down in the east through the Appalachian Mountains and in the west the Rocky Mountains. His extreme southern limit is here in the mountains of New Mexico.

Plainly he prefers coldish country with snowy winters and therefore has specialized in large well-furred sprawl-toed feet that function on the snowshoe principle and have given him his usual name. Depth of snow does not bother him since he can travel on the top layer. In fact, the deeper the snow the higher he can browse on bushes and tree twigs and bark in winter. He wears the lagomorph multiple-layer fur coat with his own variation. His reddish underfur remains the same color the year round and in summer is hidden by secondary yellowish black hairs and long black guard-hairs. In winter these latter two layers are replaced by white hairs, which blend him so neatly into snowy surroundings that even a close observer can pass by within a few feet without knowing he is there. He does his varying, his color changing, strictly by the calendar because even if kept in warm quarters with no change of diet he will do his varying right on schedule.

In recent years *L. americanus* has been of particular interest to biologists because of increasing evidence that he is particularly subject to population cycles. This is not a simultaneous rangewide phenomenon and occurs with different scheduling in different regions, but wherever it does occur, it follows an approximate nine-to-ten year pattern. His numbers will increase from year to year until a high quota is attained, to be followed by a sudden die-off. What happens apparently has little or no connection with predation or lack of food and presumably is triggered by population pressure. His endocrine balance goes off kilter, producing a slowdown of his adrenal-pituitary system. The sugar content of his blood drops and large numbers die in a condition of shock. This susceptibility to what is now known as shock disease must be a natural method of population control achieving in a more drastic and fluctuating manner what the cottontail does with territoriality.

L. arcticus, the Arctic hare, and *L. othus,* the Alaskan hare, are the two American hares correctly identified by their common names. They closely resemble each other and are the largest of the whole group in body size and obediently have the longest proportionate hindlegs. Both are extreme northerners, *othus* ranging through much of Alaska, *arcticus* inhabiting the Canadian and Greenland Arctic. Like *L. americanus* they change the color of their coats but not with the same completeness. In winter, which in their ranges lasts most of the year, they are all white except for patches of black on their ear tips. During the brief summer, when much of the ground is apt to be bare of snow, they keep the white of legs and tail and change the rest of themselves to brown or brownish grey. Some of them, those living farthest north where summers are merely milder winters, do not bother to change the color when renewing their coats and remain all white the year round.

Since theirs are snowy regions, they too have specialized in footwork. Their hindfeet do not spraddle out snowshoe-style like those of *L. americanus* (perhaps because the problem in their ranges is not so much softness of snow as crustiness of snow), but depend more on length and claw adaptations. They have extra development of the two middle digits of their hindfeet, probably for good traction on ice and snow crusts. Their forefeet are equipped with stout claws backed by strong leg muscles for digging down through crusted layers to any winter-dormant vegetation beneath. It is said their sense of smell is so acute they can detect such food through quite a covering of snow even while running along.

A companion development that helps in the food search is that their jaws have become rather long and tapering with the lagomorph incisors projecting forward, enabling them to reach into rock and ice crannies. I have heard of another peculiarity that strikes me as somewhat unlagomorphic. They are said to refuse to reproduce in captivity.

Obviously from their continued successful living under difficult conditions (so successful that apparently they too are subject to shock-disease population control to some extent), they are well adapted to their chosen habitats. But I cannot help thinking that they could have learned a better way from their little relative, the pika. With his hay-making he found a way to get along under similar and sometimes even more difficult conditions without bothering with seasonal color changes and making himself undergo anatomical changes of feet and head.

And now comes, for most of us Americans, the hare of hares, though we insist upon calling him a rabbit, seven species in North America with some forty-eight subspecies and every one unmistakably a jack rabbit. He likes wide open spaces, prairies and plains and deserts, and long ago took over all of Mexico and of the western United States, which could supply these for him. I have heard that experiments at introducing him into some of the eastern states were flat failures. Of course. Areas infested by trees, even those merely cluttered by underbrush, are not for him. They would cramp his style.

He demands room in which to operate. While he does not establish a private territory and defend it against others, he does have a fondness for a home locality within which, even when hotly pursued, he will strive to stay. But this averages out to the equivalent of about six to seven square miles — which is a lot of space for a landbound animal whose weight in pounds will not much more than match the number of miles.

His is a free and independent spirit. His only defenses against fanged and taloned and gun-toting predators are speed and maneuverability, yet he deliberately lives where he is openly exposed to predation. Obviously he believes that the unfettered life is worth the risks. The song "Don't Fence Me In" could have been written for him. I am one of those who consider no western landscape except a high mountain area complete without him. There is something somehow reassuring in the knowledge that he is there, out in his open spaces, always ready to explode into a seeming flash and a vanish, leaving a whiz in the air behind him.

There is some evidence that when his population in a given region increases to the point of overcrowding, nature imposes control by pushing to epidemic proportions the disease called tularemia, which apparently is always with him to some extent. If this is true, it is not a regular cyclical phenomenon and must be a sort of emergency substitute for what more conclusive evidence indicates is the normal natural control — in his case predation. Before we humans came along to upset the balance, the jack rabbit's population probably remained fairly constant, his efforts at increasing the amount of living matter in the world offset by the appetites and skills of his predators. This dependence on the simplest and most direct method of population control is, I believe, what has made him the most troublesome for us humans of all American lagomorphs.

What happens again and again in various sections of the west is a logical inevitable process. We humans come along in our own increasing numbers. We wage war on the predators that compete with us or that we consider dangerous and their numbers dwindle drastically. In virtually automatic response up goes the jack rabbit's population. Soon he is a downright nuisance. Worse, he cuts into our possible profits. He gobbles grass we want our cattle and sheep and horses to have. He raids our gardens and planted fields. He destroys haystacks by nibbling around the bottoms until they topple over. He becomes too successful at doing his evolutionary duty. We in turn are too impatient to see if nature will finally step in with tularemia or shock-disease control. We organize jack-rabbit drives and slaughter him by the thousands. And then, like as not, we find that the more successful we are in trimming down his numbers, the more we are encouraging the other predators we have not yet managed to eliminate, which normally feed on him, to shift their attentions to our domestic livestock.

That is a seesaw, a man-caused cycle, which still rocks along in rather regular sequence here in New Mexico. I follow it in the local newspapers. Complaints appear about too many coyotes and something must be done. Something is done. After a time items appear that jack rabbits are unusually numerous and causing damage and something must be done. Something is done. And in due time the coyote is back in the news.

I remain neutral in the matter because I feel about the coyote precisely as I have stated I do about the jack rabbit — and I am grateful that both show stubborn ability at survival. If there is any blame to be allotted from any angle on the situation, that blame belongs to us humans. The coyote and the jack rabbit worked out their relationship long before we were around to interrupt it and made it of mutual benefit. The coyote had reasonably regular meals and the jack rabbit had a reasonably effective method of population control. The coyote was the jack rabbit's device for insuring that only the best gene-carriers of his kind would survive and breed and in turn the jack rabbit was the coyote's device for transforming vegetation into meat for his meals. We humans are the ones who insist that the affairs of this world need continual rearranging to suit our particular selves.

Of the jack rabbit's seven species, the one that seems to me to be an outright exaggeration of much that is hareish in general and all that is jack-rabbity in particular is *L. alleni,* the antelope jack rabbit, largest of all the the seven, the finest runner, the most expert maneuverer, the farthest leaper — and by quite a margin the longest eared.

Like the little pika at the other end of the structural sequence he too has deliberately chosen inhospitable-seeming habitats. He lives in the hot dry wide open sun-exposed cacti-infested arid and desert area of the southwest corner of New Mexico and of southern Arizona and on down the barren dryest-of-all western edging of the Mexican mainland. He disdains the heat-escaping habit usual to other desert dwellers of retiring into burrows during the sun-drenched day. I like to think it is his pride in his adaptation to his harsh environment that gives him one of his distinctions, his obvious joy in just being alive.

Those ears, his personal specialty, must be a major asset. They have their own outsize beauty. They are big scoops of cartilage bare on the inside, covered with very fine brown hair sprinkled with black on the outside. Along the inner edges are fringes of whitish bristles and on the outer edges running up to the tips are trimmings of pure velvety white.

All jack-rabbit ears can swivel in wide arcs and are excellent sound-gatherers, operating on the same principle as does a radar scanner. They may also, as some naturalists have suggested, be effective waterproof umbrellas when their owners are caught in sudden rainstorms. Certainly the antelope jack's ears qualify as best of breed on both counts. I am fairly sure he uses them for yet another important to him purpose.

On first thought one might think that large ears would be a handicap for a desert animal. They greatly increase the overall surface
through which heat will be absorbed when the surrounding air is
above the body temperature, as it is very often in his homelands.
And his problem is how to lose heat not to gain it. Second thought
offers what I believe is the answer.

The directness and power of sunlight in desert regions creates a
definite differential between the temperature in the open and that in
the shade. Even in my own only semi-arid area, when I step from
direct sunlight into the shade of one of the trees I have planted and
zealously watered, I find that differential surprisingly distinct. It is
not only that the air in the shade is somewhat cooler than that in the
open but also that the direct radiation of the sun is no longer beating
on me. I have noted too that the temperature of the ground surface
in the shade is always some degrees lower than that of even the
shaded air over it. The antelope jack knows all this better than I
do — and takes advantage of it.

When the temperature in the open has climbed uncomfortably or
dangerously too much above his body temperature, he finds a bit of
shade — even that of just a scrawny cactus will serve him. He folds
those long hindlegs and settles down flat to the shaded cooler
ground and the contact of his body with it helps some. Those outsize ears drop down — though ready to rise and swivel about gathering sound at the slightest suggestion of danger. There in that
meagre shade close to the shaded ground those ears are no longer
heat gainers and have become heat losers and their size is all to the
good. The differential may be giving a margin of only a few degrees, but that is enough for his cooling system to function.

Another of his resources for desert living is his ability to get along
on the scant vegetation available, which supplies him with both food
and moisture. In the good seasons the few grasses and wildflowers
and rare small shrubs offer him enough of both. In the bad seasons
the grasses, retaining nutritional value even when withered and
dried, are still food. His problem then is water. He turns to some
of the varieties of cacti, gnawing into their moisture-storing inte-
riors. Apparently, too, he has in the extreme what all leporids have
to some extent, a capacity to endure prolonged periods of drought

and poor fare, then to return rapidly to well-being when conditions improve.

Antelope jack rabbit. Like all the jacks he is not a rabbit but a hare. The antelope for whom he is tagged is not an antelope but the pronghorn. His name ought to be regarded as ridiculous, and yet it is so imbedded in usage and has acquired such familiar connotations that it has earned a rightness all its own.

Like the pronghorn he has a white patch on his rump and again like the pronghorn he can turn it on and off like a signal light. It is composed of long white hairs that lie flat when he is at rest, hidden by overlying long blackish hairs. When he goes into action, the white hairs stand up, becoming fully visible, the whole patch showing as fluffy white as the little cottontail's little tail. Then he adds a trick that the pronghorn has never bothered (has not needed) to master. By flipping those white hairs from side to side he makes the white patch swing from one side of his rump to the other as he stages his version of a flash and a vanish.

In flight he invariably zigzags and with each zig and zag flips the patch over so that it is always on the rumpside closer to, more visible by, the pursuer. The true why is still a puzzle. One theory is that he hopes to confuse the pursuer into thinking the shift of direction of each zig and zag is more than it actually is. Another suggests the purpose is to fool a predator who is close enough behind to snatch at him into snapping so far to the side that the jaws close primarily on hair and slide off his rump. I lean toward the theory that claims he is advertising his identity, in effect telling the pursuer to forget it, to quit trying, to realize this is one of those things you should know by now you can't catch.

Sometimes he seems to be trying to teach that lesson to an inexperienced pursuer, perhaps to save himself bother on future occasions. He will forego the flash and vanish business and drift along just ahead of an eager chaser, doing no more than merely maintaining his lead. The effect on the morale of any but a very stupid pursuer must be drastic. No matter what frantic efforts are made, that tantalizing patch of white is always just out of reach. I have heard of dogs given such treatment who for months afterwards will not only not chase a jack rabbit, any jack rabbit, but will put up an obvious pretense of not seeing one who happens to be close by.

On the other hand in such cases the antelope jack may simply be enjoying the run and wanting to prolong it. He is as graceful in action as the pronghorn and gives every indication of liking to skim through his open spaces, dodging or sailing over upthrust cacti with obvious ease. At high speed, as in everything else, he carries to the extreme the jack-rabbit bounding gait. He seems to float effortlessly along with a rocking motion between each leap. His forefeet come down and bounce his forequarters up again with that rubber-ball effect while the rest of him is still in the air. Then his hindfeet come down about two yards ahead of where his forefeet were a fraction of a second before and he is off into another soaring leap. He is the one of his genus who can reach 45 miles per hour. His namesake, the antelope who is really the pronghorn, could beat him in a flat heat. But if he were the same size and his speed had increased in proportion, he would be at the finish line while the pronghorn was still getting started.

Since his range is restricted by his choice of habitats and offers relatively scant food resources, there have never been very many of him — in leporid terms at least. According to current reports his population is dwindling as we humans encroach ever more on his open spaces. He will not be crowded. He can usually outrun his four-footed predators and out-dodge the two-winged, but he cannot abide the advance of what we call civilization. He is retiring in his lesser numbers deeper into his deserts, where I hope he will always have a place. He belongs. To me he is another of evolution's finest achievements.

Sitting back to consider and contemplate of what I have been writing, I sense something ironic about it. I have been referring to the "amazing fertility" of the lagomorphs that sometimes results in population explosions often called plagues. Yet I am a member of a single species which nowadays probably outnumbers all of the lagomorph species combined and goes right on increasing its total.

From the perspective of other forms of life our human population explosions are the real plagues.

Now and again various lagomorphs arrive at overpopulation. I find it interesting that the North American lagomorphs (I am not familiar enough with enough of the foreign to hazard assertions about them) have developed natural methods of population control. Predation, of course, acts on all of them in varying degrees. The jack rabbits rely chiefly on that simple direct control. The snowshoe rabbit and the Arctic hares depend more upon epidemics of shock-disease. The cottontail uses territoriality with an assist from predation. The little pika avoids any particular need for control by living where the climate holds him to a very brief breeding season.

In practical results our own fertility has become more amazing than that of the lagomorphs. Long ago they responded to the possibilities inherent in their fertility by becoming subject to controls that have kept them in reasonable balance through the ages — and still would if it were not for the constant rearranging of the affairs of the earth we humans insist upon. Not yet have we humans made more than a few weak and scattered beginnings of a response to the possibilities-becoming-actualities inherent in our fertility.

I am forced to the conclusion that at this particular time in evolutionary history the lagomorphs are the more responsible citizens of this world we both inhabit.

Order: *Marsupialia*
Family: *Didelphidae*

Our One Marsupial

I DO NOT even know the man's name, but I wish him a long and joyous life. His car's tank had just been filled at the service station and he was in a hurry to depart, but he wanted to make a final point in the conversation he had been having with the station attendant. "Did you ever stop to think," he said as he ducked into his car, "that there's no bag limit on looking? Like last week I was out with a neighbor after deer. He had a license and I didn't. Just along for the outing. He got one. One deer. All he was allowed. Then he was busy dressing it and lugging it to the jeep. I went wandering on and I got five. Had a good look at each one of them too."

Which reminded me that the best hunting I ever had was one day up above Cow Creek when I sat on a cliff-top and watched a small herd of elk enjoying the sun in a smallish clearing several hundred yards down and away. I got all seven of them and I had them for almost half an hour.

I have been doing another kind of "looking" in the pages of a translation of the *Historia Naturalis* of Gaius Plinius Secundus, usually known as Pliny the Elder because he had a literary nephew of the same name. In his time, the first century A.D., wild beasts were wonderful creatures.

The *leucrotta*, for example, was "of extraordinary swiftness, the size of a wild ass, with the legs of a stag, the neck, tail, and breast of a lion, the head of a badger, a cloven hoof, the mouth slit up as far as the ears, and one continuous bone instead of teeth." It was said "this animal can imitate the human voice."

The *mantichora* was an even more formidable beast. It had "a triple row of teeth" and somehow fitted these into "the face and ears of a man." It was "the color of blood" and had the "body of a lion" with a "tail ending in a sting, like that of a scorpion." It too was of "excessive swiftness" and was "particularly fond of human flesh." It did have one gentle feature: "azure eyes."

The *catoblepas* could have been deadliest of all. Fortunately it was "sluggish of movement" and its head was so "remarkably heavy" that it (the head) was carried "bent down towards the earth." If not for the sluggishness and head-hanging, it might have "proved the destruction of the human race." Any mere human who beheld its eyes "fell dead upon the spot."

Pliny's beasts derived their wonder from the convenient circumstance that they did not need to exist. He thought they did. But his "knowledge" of them came primarily from folklore and legend and myth.

Nowadays amateur naturalists like me feel a compulsion to stay rather close to those cold hard things called facts. We are bound to real beasts — to what we think we know about them from personal observation and experience and to data about them assembled by the whole wide expanding array of the biological sciences. A dull business? Not at all. As that kind of knowledge increases, we discover that fact defeats fiction, that our beasts are more wondrous than those of the Plinys of the past. They can fill us with infinitely more wonder at the magic and mystery and intricate intertwined kinships of this spark of being called life we share with them.

Consider one of the homeliest, scraggliest — and wonder-full-est — of all mammals. *Didelphis marsupialis,* simply *the* opossum.

As her scientific name indicates, she has ——

As that beginning indicates, almost automatically I think of the opossum in feminine terms for what seem to me the very good reasons that the female, not the male, has the natural equipment that distinguishes the whole marsupial order from the rest of the mammalian orders and that the most fascinating aspects of the species have to do with maternal, not paternal, matters. Perhaps I am influenced, too, by the fact that my most vivid and memorable meeting with an opossum was with a female surprised and blinking in the sudden glow from my flashlight and doing what no male ever does — carrying nine opossum youngsters in a tangled group on her back . . . That is not quite accurate. Two were riding her tail.

As her name indicates, she has remarkable peculiarities. Genus name from the Greek *di-delphis,* double uterus or womb. Species name from the Latin *marsuppium,* purse or pouch. She is a mammal who has a double womb and carries a pouch about with her as part of her natural endowment.

Right now let me dispose of her mate. Only one thing about him, as separate from her, I regard as of real interest and even that is designed to serve her. To match her double womb he has a bifurcate penis, a double or forked penis. That said, he can be pushed aside. He leads a rather solitary life, associating with her only when she is in estrus, which normally occurs every twenty-eight days — if she did not become pregnant the last time. In all other respects opossum family life is completely hers.

The opossum belongs to the order Marsupialia and is the only member of that order in the whole of the United States. I have heard that in 1917 two representatives of another branch of her family, an Alston's mouse opossum and a Mexican mouse opossum, were found in the New Orleans area of Louisiana. Beyond doubt they had arrived there as stowaways aboard a banana boat from Central America. In 1935 another Alston's mouse opossum was discovered near New Orleans. Obviously another surreptitious entry by way of passage with a cargo of bananas. I suppose that if a sufficient number of such visitors reached our southern shores, they might try to establish a colonial foothold. I say "try" because I believe they could not succeed in any such enterprise. They are tropical animals who need the kind of diet the tropics provide. Unlike them, *Didelphis marsupialis* was a confirmed resident of what is now the United States long before there was a United States. To identify her properly I have to go well back into the taxonomic tangle.

All of us mammals belong to the one great class, Mammalia. The experts can cite other items, but two things really distinguish us all: we have hair at some stage in our existences and our females suckle our young, have mammary glands for that purpose.

All of us mammals in the Americas also belong to the subclass Theria, beasts who bear their young alive. That subclass distinction has to be made because of two creatures who came far enough along the mammalian road to have those two distinguishing features, then held fast to another inherited from our reptilian ancestors. They insist upon laying eggs. The platypus and the spiny anteater. Both Australians. So the class Mammalia has to be split into two subclasses, Prototheria for the odd two; Theria for all the rest of us.

That settled, the opossum and her marsupial relatives present another difficulty. They came further along the mammalian road but not quite all the way. They got rid of the egg-laying business, probably figuring as do the rest of us that method is for the birds, but they did not evolve or only partially evolved the final mammalian reproductive device, the placenta, the enveloping membranous organ that develops in the female during pregnancy to protect and nurture the growing embryo. They opted long ago for the pouch method. So our subclass Theria has to be split into two infraclasses, Metatheria for the marsupials, Eutheria for all the rest of us multitudinous placental mammals.

What intrigues me as much as anything else about our opossum is that right now she is demonstrating one of the processes I read about in books on evolution. She is expanding her range as a result of changes in environment.

Way back in the very beginnings of this age of mammals marsupials were rather widespread around the world. The probable common ancestor of most or all of them was a small opossumlike animal not much different from our contemporary opposum. Then we placentals came to the fore and began to crowd the marsupials out. They finally disappeared everywhere except in Australia and South America — two land masses isolated from the other major land masses. In those two continents with no new and more efficient placentals able to immigrate and offer competition, they went ahead to evolve all manner of genera and species roughly paralleling those of us placentals elsewhere. Such things as marsupial mice and rats and cats and wolves and bears. Among them in South America a populous family of many kinds of opossums.

At last the Central American land bridge was re-established, becoming what evolutionists call a faunal filter between South and North America. As could be expected, we pushy placentals began filtering through southward, crowding out some of those backward marsupials, a process still going on.

One group of the South American marsupials started filtering northward. The opossums. Quite a variety of them. All but one of them have since been content to come no further than Central America or a short way into Mexico. The adventurous exception is *Didelphis marsupialis.*

She came up out of Central America into Mexico, working her way along the coastal areas and somewhat inland. She did better along the Gulf coast and wandered on into what is now Texas. Aha! Rather good opossum country on eastward through what are now the deep south states. By time Columbus launched the European human invasion of the Americas, she was a well-established North as well as South American, ranging through what are now the eastern-southern states and moving further north all the time.

There had been no marsupials in Europe for many millions of years. The Europeans coming here had no experience with such a beast. When they met her, she seemed straight out of a Pliny-style bestiary. During the next century or two she was probably the subject of more written accounts, most of them nonsense, of more outlandish names, of more wildly imaginative drawings, than any other American animal. For a single example, Topsell following Gesner, called her the "SIMIVULPA, or Apish-Fox" — a creature "in the forepart like a Fox, and in the hinder part like an Ape."

Of all the ridiculous notions about her I have encountered the most curiously ridiculous is one which still held its vogue among the gullible well into this supposedly enlightened century. Apparently it started as an attempt to solve the seeming riddle of the male's forked penis. Such a sexual instrument, so logic seemed to dictate, would require double orifices in her for him to be able to deposit his sperm. Only in one part of her anatomy was there a pair that could qualify. So the notion was launched that he inseminated her by depositing his sperm in her nostrils. That posed the following problem of transference of the sperm to a more congenial maternal site. So the corollary was added that she would immediately thereafter poke her nose into her pouch and indulge in a hearty sneeze. Believe it or not, I have read a report of that one still having its believers in the 1950s.

Most of the early scrambled descriptions were based on ever more elastic accounts of one opossum who reached Europe very early. The captain of Columbus's Niña, Vincente Yañez Pinzón, on an expedition of his own in 1500, met her in Brazil and took one of her back to Spain to be presented at the court of Ferdinand and Isabella. But it was Captain John Smith of Jamestown repute who eventually gave her what was to become her lasting popular name in English. According to the perennial tale, he asked an Indian what this strange beast was called. The reply sounded something like "passum." The word was preceded by an Indian grunt or belch. So Captain John, when he wrote his *True Relation* in 1608, made it "opassum." The "a" probably became another "o" through the medium of the now traditional American sloppiness of diction.

During these last few centuries while we humans have been busy spreading out through what is now the United States, the opossum has been doing some of the same. In a few instances she has had direct assistance with the best example of that provided on the west coast. Something over half a century ago she was imported on a small scale into Southern California and a corner of Oregon. Since then she has spread up and down the coast with a range there now extending from up in Canada down the whole length of California and into western Mexico. And meantime on her own she has covered the whole eastern half of the United States except for the final upper reaches of New England, has come west on a large front into Colorado and New Mexico, and has meandered up into Wyoming. I would take odds that by the end of this century she will have covered virtually all of the contiguous United States except for the true desert regions.

Her getting about is not really so remarkable in terms of territory covered. Any wandering small beast not afflicted with territorial ties and equipped to carry her young around with her should be able to colonize quite a lot of area in a relatively short time. More remarkable is that this particular little beast does it despite such apparent handicaps.

She seems to be wonderfully stupid. Her brain does have well developed sense centers (smell and touch, for example) and those regulating breathing and heartbeat and blood pressure and other vital functions. But that brain is lamentably deficient in frontal lobes, the portion associated with intelligence. And the whole is small. Not just small because she is small, but small in proportion to the rest of her. Bailey once filled with beans the brain case of an opossum and another of a raccoon, a placental mammal of about the same skull-size and overall size. The opossum's held twenty-one beans, the raccoon's one hundred fifty.

She is usually, for such a scraggly beast, rather well fleshed and endowed with fat. Many humans, particularly those of southern states background, insist that when associated with sweet potatoes she is definitely edible. Which notion (even without sweet potatoes) is shared by just about any and all four-footed predators of her own size on up.

She has very little of defensive armament to help her. She can run, but not in a class with most predators. Her claws are of scant use in a fight. She does have fifty teeth and as a result her grin can have a frightening raggedy look, but those teeth are not much as teeth go and she must know that because she is not much of a biter except of what she regards as food. Though she has vocal equipment, she does not even do much with it except on occasion for a kind of ridiculous low hiss or a silly muted growl. As a battler she is a pushover. Even her eyesight is poor. Any ordinarily bright light blinds her. Average sunshiny daylight can do it.

How does such a backward, poorly equipped excuse for a member of our splendid mammalian class manage to get along, let alone get about? How has she managed to do the seeming impossible, make her way up through the Central American faunal filter against the current of supposedly more efficient placental mammals going the other way? How does she manage to keep on extending her range through this North America supposedly ruled by placentals?

Well, now, we humans have been helping her some. As noted above, we have given her a start in some areas by taking her to them. We have helped her with our dams and reservoirs and artificial ponds and lakes and even with our watering tanks for stock. She appreciates these because she prefers regions in which dependable water supplies are available. But the major manner in which we have helped her is by our being the most efficient of all predators. In region after region we have exterminated some and drastically reduced many more of the other predators who, dining often on her, would have hampered her colonizing activities.

None of that offers more than a very partial explanation. She had very little if any help from us when coming up northward and taking over chunks of Mexico and spreading into what is now the United States to be ready to greet Captain John Smith at Jamestown. She did it on her own. How?

She has, to begin, what biologists agree is an asset for survival, a "general lack of fastidiousness." That has nothing to do with tidiness. She has little in line of looks, but she takes care of what she has. Like a cat, she frequently grooms herself. Her lack of fastidiousness applies to more basic matters, such things as food and shelter.

She is virtually omnivorous, eats anything eatable. No patent on that, of course. We humans, for example, are rather good at it. She is better. Put her at the opposite extreme from say, the koala bear (another marsupial) who is so fastidious he eats nothing but eucalyptus leaves. She will dine as occasion offers on insects, worms, reptiles, fish, small mammals, birds, eggs, mushrooms, snails, seeds, vegetables, fruits, grains, and more. She is not averse to any carrion she can find. A partly filled garbage pail with the lid off would be a whole cafeteria to her.

Incidentally, her appetite for insects should warm our hearts. She may actually eat more insects in winter when few are about than in summer when they are abundant. In winter her diet is otherwise more limited, so she searches out, in hollow logs and ground holes and such, insects that are hibernating, waiting for spring to attack man's crops. Her efforts in that respect are to be preferred to the use of DDT and other insecticides.

She needs a den as a hide-out, as a protection against daylight glare, as a shelter from cold in winter, as a safe place for the very brief period during which she gives birth to her young — who immediately thereafter will be going about in her pouch with her. No fastidiousness here either. Any place that meets those requirements will do. She will set up her skimpy housekeeping in abandoned burrows, in small caves or rock fissures, in hollow trees, in piles of trash or other debris, in vine tangles, in old squirrel nests, underneath sheds and barns, in empty houses. Not being a fussy tenant, she can find a temporary home (all she wants) almost anywhere she goes.

Another of her ways is to compensate for her lacks.

I said her eyesight is poor. Her eyes seem to be all pupil, the irises showing only when she is in bright light and the pupils are contracting against it. Obviously those eyes can function only in dimness or the dark of night. Therefore she sensibly is nocturnal in habits — and even then depends more on her senses of smell and touch and hearing. She makes the most of those centers in her tiny brain that are rather well developed. Not like us, who let such centers dwindle in neglect or only partial use while we put such reliance on our optic and frontal lobes.

Since she is a poor fighter and not much better as a runner, she has made herself a very good climber. Her toes are more flexible than those of the feline and canine variety. The five digits on each front foot have sharp claws like those of the squirrel. The five on each hind foot are really remarkable. Four have claws — and the fifth is what can only be described as a clawless opposable thumb. Her hind feet resemble human hands. Of all living things, only we humans and the apes share that efficient grasping device with her.

Her tail, one of the ugliest things about her, longish, ratlike, almost hairless, dark about half its length and yellowish white to pink the rest of the way, is not quite the "extra hand" some writers have labeled it. But it is a real help in climbing, being prehensile, able to curl about and grasp things. When young she can hang from a tree branch by it, a trick that becomes more difficult as she puts on the weight of maturity. At times she has been seen using that tail to carry nesting material.

Climbing is a defense against ground-bound predators, not

against other climbers. It is of no use at all if she is jumped too far from something to climb. So she has developed another type of defense that sometimes works, what is mistakenly known as "playing possum." She is not playing, feigning, putting on an act. Her whole system is responding in a manner that is involuntary, in a sense automatic. For a few seconds her heartbeat and breathing speed up and she loses control of some bodily functions. Then she goes, almost suddenly, into a trancelike catatonic state in which she seems to be dead. Heartbeat and breathing are virtually imperceptible. She can be lifted, tossed about, bitten, have her limbs bent painfully, her whiskers pulled, even her eyeballs touched — and there will be no reaction.

Again no patent on that. Similar behavior is sometimes seen in other animals, even on rare occasions briefly in us humans. But the opossum has developed it into a positive defensive ability easily triggered. As I said, it sometimes works — when she is up against an enemy not interested in immediately eating her.

Even when it works it often has a disadvantage that has required her to evolve another ability, a special kind of vitality, an amazing capacity to recover from cuts and scratches and bruises and to repair broken bones. Being particularly vulnerable to injuries when "playing possum," she can collect many of them. And then recover. I know of an opossum who, as an autopsy revealed, had at various times recovered from two broken shoulder blades, eleven broken ribs (two of these broken three times, two others two times) and a badly damaged backbone.

I come now to what must be her major method of survival and of extending her range — her reproductive facility. In this she is not quite a match for the leporid lagomorphs, the rabbits and the hares, who surpass her by breeding more often and thus producing batches of young more frequently than she does. But in her own way she does remarkably well — and sets some records of her own.

At mating time she produces internally an average of twenty-two eggs. With the double sex-equipment she shares with her lazy loafer mate, sometimes all of those eggs are fertilized. Immediately the embryos begin to develop. Just twelve days and some hours later they will be born. In not quite thirteen days each embryo has become a creature with air-filled lungs and functioning digestive tract and neuromuscular system. In other words it is breathing and capable of eating and moving about.

Opossums are not very big. Even so, when born, the young are unbelievably tiny — the whole litter would fit in a teaspoon. Yet their first experience of the outside world is a confrontation with a tough problem. They must get as quickly as possible to the warmth and security and free lunch counter of their mother's pouch. They must do that on their own. It is an upward journey through a jungle of hair, for creatures so tiny a long one. Once in the pouch they must still hunt about through another jungle to find the nipples. Their mother has licked them clean of embryonic fluid as they emerged and a little earlier, cleaning her pouch for the reception, has moistened a pathway for them to follow by licking through her fur. Beyond that she gives them no help at all.

She does not need to help them because these infinitesimal infants already have sufficiently well developed the two portions of their anatomies necessary for the start of their outside-the-womb existence. First, a pair of tiny forelegs with teensy-tinesy claws for the climbing expedition. Second, an effective mouth and adjoining muscles for fastening to and nursing-from the nipples inside the pouch. So effective is the fastening equipment that once they have found the nipples it would be difficult for weeks thereafter to detach them.

Another factor enters right here at the start. Their mother has thirteen nipples in that pouch — a number odd in more ways than one. Yet frequently there are more than thirteen young ones in the litter — sometimes, no doubt rarely but sometimes, up to the possible twenty-two. The extras are expendable, doomed to quick starvation. Natural selection applies with special emphasis to infant opossums in the very first moments following birth.

The thirteen lucky ones (not always that many make it but the score is usually good) slow down some in growth in the pouch but still do very well. After about sixty days they are the size of small rats, well furred, already almost complete replicas of their mother, able to run about and climb and ride along with their mother by clinging to her soft fur and longer stiffer guard hairs.

In a brief time they will be able to fend for themselves. When mature (a matter of only a few months) they will have increased their birth-weight at least 8,400 times. Which interesting figure is obtained by dividing the average birth weight, one 175th of an ounce, into the average adult weight, three pounds. But sometimes a plump elderly (say two years old) opossum will reach six pounds. That doubles the increase to 16,800 times. We humans average the ignoble figure of a mere 20 times.

I call her stupid and she is. Yet, on the evidence assembled by the paleontologists, I must, and I do it willingly, also call her an amazingly efficient organism. Quite possibly a much more durable one that what I myself represent.

Her ancestors evolved in the Cretaceous period, something over one hundred million years ago. Some of their descendants pushed off on their own to become the other marsupials of South America. She and her closest relatives, particularly she herself, held more strictly to the original marsupial model. It apparently is a plain fact that she has been just about exactly what she is now for at least 70 million years. A record no other mammal, not even the other members of her order, can match. How long has *Homo sapiens* been around? Perhaps five hundred thousand years and our immediate hominid ancestors perhaps two million years.

As I consider and contemplate of that scraggly little beast, the opossum, I feel twinges of a kind of awe — and of nagging doubts about my own kind. Long long before evolution even started to aim discernible experiments in my direction, she was here doing well on this earth we now share with her. She is still here and still doing well. She suggests to me that neither size and strength nor brain power is necessarily the answer to long-continued survival. She saw the dinosaurs go. Will she see my kind go too?

Order: *Rodentia*
Family: *Geomyidae*

Our Hardest Worker

NOBODY loves a gopher. Not even another gopher.

That is his own fault — or, perhaps, from his point of view it is a virtue not a fault. He does not want to be loved. He wants only to be left alone to live out his life in his own way. And his way is to sentence himself to solitary confinement and hard labor for as long as he shall live.

He is a confirmed grouch, always bad tempered, who lives alone by deliberate choice and refuses to tolerate visitors even of his own kind. When two gophers meet, which is a rare occurrence, they fight vigorously, trying to get a grip on each other's snouts, and sometimes the loser loses not only the fight but his life too.

There are a few certain people who might admit to a small spot of affection for him: scientists who have earned degrees or added to their publication lists by doing research on him and publishing the results. Even so, the sum of their scientific writings inevitably emphasizes two items: that he is definitely antisocial and that he offers very few if any endearing habits or qualities. Even more damning, almost all other kinds of writing about him are almost completely devoted to attacking him as a pest, to prescribing possible methods of destroying him, and to describing frustrations attendant upon attempts to destroy him.

I call him simply the gopher for convenience only. He is really the pocket gopher and deserves the full title. His major distinguishing feature is a pair of pockets that he wears with fine disregard for facial beauty on his cheeks. They have openings independent of, outside of but close to, his mouth and extend back along his cheeks to his shoulders — which is not as far back as that seems to indicate because he has practically no neck. Since those pockets or pouches can be considerably distended, they have surprising capacity. Shrewdly placed protractor muscles enable him to pull the rears of the pockets forward to empty them, then back to normal position. Sets of sphincter muscles enable him to close the openings as if they had drawstrings.

I have never had (I have never wanted to have) a chance to inspect a dead gopher and only a fool would try poking a finger into a pocket of a live one. Nonetheless I used to wonder about the insides of those interesting receptacles. All accounts I could find prattled learnedly about position and musculature and such. How about such things as texture and color and shape? Then I came on Volume V, the one on zoology, of the many-volumed government *Report upon Geographical and Geological Explorations and Surveys West of the One Hundredth Meridian,* published in 1875. In the section on mammals Drs. Elliot Coues and H. C. Yarrow satisfied my curiosity. Speaking to me across a century they told me that the short soft fur of the lining of the pockets is "nearly pure white." And they supplied a truly fascinating item: a pocket, when "fully everted," is "squarish or rather trapezoidal in shape, with decided corners, like a small pillow-case."

Precisely the shape we humans prefer for our own pockets.

Since he is (and always has been) strictly an American, the gopher was something new to the early settlers from Europe. Since he is (and always has been) primarily a Westerner with only a minor branch of his family found east of the Mississippi and that only in the deep south, he was not among the first new beasts to be met and named by the newcomers. As a result he had some difficulty winning the name of gopher.

It is derived, so the experts seem to agree, from the French word for honeycomb, *gaufre*. The corruption or derivation, gopher, was originally applied in this country to a tortoise who happens to be the one true member of the tortoise family native to North America, genus still known as *Gopherus*. The *Gopherus* tortoises had (you could argue still have) some claim to the name because they do a mildly impressive job for such lethargic creatures at honeycombing the ground with their burrows. But they fade into insignificance alongside the gopher. He is, without any doubt, the most active, the most industrious, the most efficient, the most all-around-complete honeycombing burrower in the whole of the animal kingdom. When his prowess in that respect became known, he clinched his claim to the gopher name.

He belongs to the order Rodentia, the rodents, from the Latin *rodere*, to gnaw — that amazingly successful order of usually small to smaller beasts. Most rodents engage in some burrowing activity, some downright expert at it, but the gopher has moved far beyond mere expertness. He has made of burrowing a business, a profession, a science, an art. It is his reason for being, his lifework.

He is the one rodent who is completely fossorial in life-style: that is, completely adapted to digging and to living underground and to spending virtually his entire life there. The mole is his only rival in this among North American mammals and the mole is an insectivore, not a rodent. Moreover there is a distinct difference in their respective burrowing techniques. The mole does little real digging; he pushes his way along just under the surface of the ground, disposing of the dirt he displaces by humping some of it above him and pressing the rest to the sides and down. He merely shifts its location slightly. The gopher is a true engineer, an excavator who works usually from six inches to a foot underground (deeper in soils subject to cave-ins) and who digs out the dirt and transports it to the surface to deposit it in characteristic heaps or mounds at the upper endings of short lateral tunnels along the courses of his main tunnels. His way involves much more and harder work and sometimes I suspect that is its attraction for him.

As an insectivore and for his type of burrowing, the mole prefers relatively humid regions where the ground is apt to be easily worked and to nurture a good supply of his kind of food, insects and earthworms and such. As a rodential vegetarian and for his more ambitious type of burrowing, the gopher prefers regions where the ground is well drained or semiarid or even arid but still supports enough plant life for his needs. Such support can be very scanty because he is so dedicated to hard work that he is willing to labor prodigiously for his meals. Thus the two of them, mole and gopher, the two American fossorial animals, are not competitors. Differing in dietary preference and tunneling technique, they have roughly divided North America between them with little overlapping of ranges.

Of the two the gopher is much wider ranging and more willing to take on touch conditions. He finds homesites in much of Canada, through all the western half of the United States plus that tortoise-shared area in the deep southeast, and right on down southward through most of Mexico and much of Central America. Prairies and plains and mountains and even deserts are all the same to him as long as hard work can make a living there. It is a safe bet that thousands upon thousands of gophers are at work right now, this very moment no matter when you are reading this, in locations from below sea level (as in the Salton Sink) on up into mountain heights even above timberline. He will get along where a mole would never conceive of trying. But, of course, he is willing, he seems to want, to work harder. He does practically nothing else but work. He is afflicted or blessed in extreme degree with what is known among us humans as the Protestant work ethic.

Long ago evolution designed him for digging and did such a good job that through the millions of years since (twenty and more) he has changed very little if at all. He is a thickset round-bodied muscular little beast averaging about eleven inches in length with three of them rather thick tail. His skull is longish and low — even among rodents he is a low-brow. His seeming lack of neck is due to a thickening of bones in that region and the attachment of well developed shoulder muscles. His legs are short and even so he carries himself in a crouching position. His feet are five-digited with strong claws, particularly his front or digging pair. In fact those front claws are so long that he folds them under against the soles of his front feet when merely walking, not working.

His body fur (he has no undercoat) is soft and oily-greasy and by some alchemy is dirt repellant so that even moist soil does not cling to it. His eyes are small with lids that can be tightly closed. Specially developed tear-glands secrete a thickish fluid that easily cleans away any dirt that gets past those lids. His ears are so tiny they can scarcely be made out through the fur around them and they have inside valves to close the openings when he is busy burrowing. In overall outline he resembles a small projectile with virtually no protrusions to hamper fast movement along his tunnels.

He does not even have to emerge from those tunnels to find water. His vegetative food supplies him with all he needs.

Not even his specialized forefeet can do all his digging for him. His teeth do the tougher work. They offer one of the best examples anywhere of the distinguishing specialty of all rodents.

From the beginning of the order the rodents, determined to be gnawers, concentrated on their front teeth, the incisors, reducing the number to just two above and two below and enlarging them. Then they held on to their back teeth, the grinders, and eliminated the others in between. Thus the common pattern is the two pairs of large incisors with gaps along the jawlines back to the molars. Their lower jaws are so hinged that these can be moved forward to bring the cutting edges of the incisors together and back to let the incisors slide past each other while the molars come together. Then the lower jaws can also be moved sideways for the grinding action. I can perform in some degree the same maneuvers with my own lower jaw — but not in any class with the rodent professionals.

Since they are gnawers and their incisors need constant sharpening to retain keenness, they are constantly being worn down. To offset this they are set in an open pulp cavity and grow steadily throughout life. In consequence they must be used well and frequently or they will grow beyond the proper useful length — and keep on growing. A rodent who neglected his gnawing would find his front dental armament becoming a handicap, in time such an obstacle even to eating that he would starve to death.

Wearing down is not enough in itself to keep the cutting edges sharp. Along the front edge of each incisor is a layer of extra hard enamel. The wearing differential between this layer and the rest of the tooth maintains a slant end with keen edge — precisely what a good carpenter does with his chisels.

The gopher has rodent incisors seemingly enormous in proportion to the rest of him. He may be proud of them because he never hides them. To be exact, he cannot hide them. When he closes his mouth his lips fold in *behind* them. That is a neat arrangement for burrowing purposes, enabling him to use his twin sets of chisels without getting mouthfuls of dirt, but it is no aid to beauty. Perhaps the fact that such a facial array is so contrary to the usual is what makes it repugnant to us otherwise endowed, but in a head-on close-up look he certainly is not pretty. If his cheek pockets happen to be well filled, bulging out at the sides of those so-visible big incisors, he is downright ugly.

Perhaps he really is proud of those incisors; in practical terms he has reason to be. They are efficient tools and he uses them efficiently. So efficiently that he can cause trouble for us humans if, in his endless burrowing, he comes on tree roots or such things as underground wires or cables. He is apt to gnaw right through whatever is in his way. He can ruin an orchard with ease and keep utility company repair crews busy.

Those incisors could be responsible for his obsession with work. They grow so fast he could scarcely consider a vacation from working — from wearing them down. He holds even the rodential record in this, which may be the record for the entire animal kingdom. His upper incisors grow about nine inches a year, his lower an astounding fourteen.

And yet, the reverse could just as well be true. His work obsession could have come first and evolution have obligingly equipped him with tools to match it.

His burrowing lifework has two purposes: to construct and maintain the extensive home he fancies and to find and keep on hand a good supply of food.

His home, usually several feet down, may have as many as ten rooms with interconnecting tunnels, the majority of rooms pantries that he tries to keep well stocked because he is a worrier and frequently has more food in storage than he will ever eat. If some of it begins to spoil, he simply closes off that pantry and digs another. One room, comparatively spacious, will be his bedroom, and in the case of females, another will be a nursery. Yet another will be the toilet chamber — and when it becomes a bit bothersome in terms of odor or space, again he simply closes it off and digs another.

But preparing and maintaining home quarters are mere minor labors for him. It is when searching for food that he really settles down to work. Since his foraging is done underground, he has to burrow his way along, the energy he puts into it almost defying belief. He burns up energy so fast that he has to fuel his internal engine with at least half his own weight of food per day. He thinks nothing of burrowing hundreds of feet in one work-shift. Researchers trying to assess the scope of his activities have dug up half a mile of tunnel before giving up the project. Audubon and Bachman once became curious about some gopher mounds in a friend's garden. Being lazy (ungopherlike) they summoned "several servants" to do the digging. Soon "galleries" were uncovered going here and there (honeycombing), some dropping deeper to pass under thick paved walks. Then the discovery was made that others led out of the garden and across fields and into a woods beyond. And so that project was abandoned. One gopher, probably only one digging day ahead of them, had defeated them.

Biologist Victor H. Cahalane has calculated that a one hundred fifty-pound man would have to dig a trench seventeen inches square and seven miles long in ten hours to match what a one-pound gopher often does in one night's foraging.

His earth-moving method is all his own. Down there in his clean-cut almost cylindrical tunnel, which will be two to three inches in diameter depending upon his personal girth, he spreads his hind feet to brace them for leverage and digs in ahead with fast strong stokes of his forepaws. Loosened dirt is swept back under his belly and at intervals he uses his hind feet to kick it back further. If he encounters hard-packed dirt or some obstacle, those ever-growing incisors come into play. When he has accumulated a fair pile of debris behind him, he executes his own brand of somersault, pokes his nose under and back between his legs and turns himself over with a twisting motion that puts him on his feet with directions reversed. He is now facing the pile of debris. Contrary to popular notion, he does not use his cheek-pockets to carry any dirt. They are reserved for food and nesting material. What he does do is turn himself into a miniature bulldozer. He drops down on his chest, puts his forepaws up flanking his face with the long claws extended fanwise, and starts pushing with his hindfeet. He scoops up the pile of debris and shoves it along ahead of him. He has found that the best procedure is to progress in a series of spurts, a few inches at a time, pulling back a bit at each interval to regather any dirt that has tried to slip past him.

Back along the tunnel goes that load until he reaches the lateral disposal tunnel currently in use. Up this goes the load and a final flip deposits it outside. When he thinks that a sufficient mound has been created or, more likely, that the pushing distance is now so long that a new lateral is in order, he seals the opening of the one to be abandoned and digs another. This sealing off practice, very sensible to discourage unwelcome visitors, sometimes presents a mystery to people unacquainted with his habits. They suddenly come upon mounds of fresh-dug dirt where most certainly were none a short while before and are unable to discern any reason or source for them.

Since he lives underground, night and day are much the same to him. He is apt to be working any time during the full twenty-four-hour cycle and when he works, he WORKS! I am a bit too big to crawl down into one of his tunnels, so I have never seen him at his actual digging. But I have watched little loads of dirt come popping out of a lateral — usually without even so much as his head coming into view. They come at a surprisingly rapid rate. Out pops a load. Now he has to go back to the tunnel on which he is working, excavate some more, execute his somersault, gather the debris and bulldoze it to the opening. In about the time it takes merely to think through that sequence — aha! there comes another load!

Perhaps a longer interval occurs. He probably has found what he is looking for. A juicy root, a tuber, a bulb, an underground stem. He is cutting it into pieces and stuffing it into his pockets. That will not take long. His forepaws can move so fast that if he could be seen all that would be seen of them would be a blur. Some naturalists claim they can move so fast they produce a buzzing sound. But look! Another load is coming out. He has not bothered to take time to deposit that food in a pantry. No need. It is safe in his pockets and he will not stop working to empty them until they are as full as he can get them.

He works not only anytime around the clock but also around the seasons. No hibernation for him — which is, after all, a means for those who practice it of getting through a time of year when food is very difficult to find. Working underground, he can find it in winter as well as in the rest of the year. He can burrow along below the freeze level or, if that is unproductive, he can attack even frozen ground with those chisel teeth. Heavy snows, of course, can be a nuisance, blocking normal use of disposal tunnels. He solves that by making tunnels in the snow along the surface of the ground and pushes his burrowing debris into them — leaving when the snow melts roughly cylindrical deposits sometimes known as earth-sausages and to the knowing as gopher-sausages. Even in mountain heights he can survive cold weather so well that he often weighs more at the end of winter than he did at the start.

I write that he is fossorial, spends his life underground. There are just two things that induce him to emerge above ground and these are precisely what I would expect them to be: food and sex.

Being such a worrier, having such desire to keep his pantries full, he is occasionally tempted to harvest some of the succulent green growing above ground. The wise gopher uses a safe technique — which in its turn presents another mystery to unknowing observers. He takes hold of a plant by its roots and pulls it down into his tunnel. A wanderer in a gopher locale may see a small plant begin to quiver and shake for no discernible reason, then sink out of sight into the ground. But many another gopher, when the urge for fresh green salad overcomes him, is more reckless. He pokes his head out of a burrow (usually but not always at night) and reaches around to dò his harvesting with his hindquarters still in the hole ready to draw him back down in a hurry. If temptation is really strong and he is somewhat adventurous, he may actually move a few feet out into the open. But only very briefly and with every motion of cutting his crop and storing it in his pockets done with speed and precision. He knows very well that all manner of furred and feathered and scaled predators regard him as definitely edible.

I know no valid reason other than his cantankerous solitariness why his sex life could not be as subterranean as the rest of his existence. Since he insists on living alone, when he does desire dalliance with a female, he has to go looking for her. He becomes restless, even more grouchy than usual, at last is driven to dare the upper world and go searching. How he finds her, since she is as much an underground recluse as he is, has not yet been reported. She may have emerged too and be waiting for him to come along, though it is likely she will have stayed very close to her home and the nursery she may already have prepared. Again, it is a fair guess she will simply be doing the gopher thing, burrowing, and he will detect her presence by means of his unusual sensitivity to ground vibrations, determine she is a female and in a receptive mood by his very effective nose, and either find her burrow entrance or dig in to her.

Their romance, if it can be called that, will be brief and, insofar as observation of captive couples has shown, not ecstatic, merely matter-of-fact. Her attitude seems to be chiefly that of submission, though she may appreciate the affair more than he does. At least she has been known to emit a mild purring sound while he remains silent, seeming to regard mating as an obligation to be performed and done with. Then he will depart, leaving all consequences to her, and either return to his bachelor quarters or, if that be blocked or inadvisable for any reason, simply pick a new site and start work on a new home and ever-expanding tunnel labyrinth.

The mortality rate among gophers is relatively high, and the female has a job to do maintaining population figures. All evidence indicates she does well. She reaches puberty at six to nine months, has a gestation period of twenty-nine days, averages four infants at a birth, and often submits to him sufficiently to have two batches per year.

About all that is known of what goes on down there in her underground home when the nursery is occupied is that the young are born blind, hairless, toothless — and pocketless — and she must feed them well because they grow rapidly. Does she do more than feed them and keep the nursery clean? Does she give them lessons in burrowing? Probably not, at least not deliberately, simply by example. Being a gopher, she probably is using all her available time justifying her existence by digging away and stocking pantries. I can imagine them, when they can move about, exploring her tunnels, following her, watching her work, perhaps trying some digging themselves — and scurrying out of her way when she executes a twisting somersault and bulldozes a load toward a lateral opening. But not for long. By time they are three to four months old, they realize they are gophers too, the gopher ethos asserts itself, and they depart to mark out their own solitary underground land claims.

In the foregoing I have been writing about a gopher who does not exist, a generalized gopher with his characteristics as accurate as my knowledge permits but with physical facts about him based on averages for all gopherdom. The true gopher is many gophers. Since his overall range is so large and includes such varied territories and climatic conditions and since localized gopher populations are so restricted by the underground life, he has developed many degrees of differences in his far-scattered habitats. According to what portion of the range he inhabits he varies in size, in coloration, in length and development of claws and teeth. Though he is always brown, for example, he can vary from very light tan to dark sepia, almost black, and sometimes have a few whitish spots on his head. But always and forever, wherever he is, he is unmistakably a gopher.

Yes, he is always a gopher and his antisocial attitude might be said to extend even to his taxonomic family. No cousins admitted. Gophers only. Although the taxonomists are still fussing with the family listings and although they have managed to cite eight or nine genera, at least thirty-nine species, and more than four hundred subspecies, one fact is certain: all are distinctly gophers.

Order Rodentia. Suborder Sciuromorpha, the squirrellike rodents, this too from the Latin, *sciurus,* squirrel. Family Geomyidae, which my limited scholarship suggests is misleading since it is derived from Greek roots adding up to earth mouse. He is not a mouse. The true mice and rats are the leaders of a whole different suborder of rodents.

While the experts still disagree on the number of gopher genera, they are unanimous that three genera are citizens of the United States. There is one simple way to tell the three apart — that is, if whichever gopher you are trying to identify lets you have a close look at his protruding incisors. *Geomys* has two longitudinal grooves on the front surface of each upper incisor. *Pappogeomys* has one groove. *Thomomys* has none.

Ungrooved *Thomomys,* logically known as the smooth-toothed pocket gopher, is the widest spread and most numerous of the American three — in fact of all gophers anywhere. This country has no monopoly on him because he is a Canadian and a Mexican too, but we have by far the largest portion of his population. He can claim six species, no record in gopherdom, but he is way out ahead of all gophers in number of subspecies with a total approaching three hundred. With all the minor adaptations represented by those species and subspecies he has no difficulty adjusting to a variety of environments. He provides ample proof here in New Mexico.

He lives here in desert areas where plant life is sparse and composed chiefly of agaves and cacti and yuccas, all wonderfully protected above ground by saw-teeth and spines and positive bayonets. Like a land submarine he attacks these from below and harvests from the roots right up into the insides of the outwardly armed plants. He also lives here on up through the mountain areas where conditions are quite different, even doing some of his digging on the very top of Wheeler Peak, highest point in the state. I like to think of the pika up there, serene on a high perch, occasionally catching a glimpse of gopher activity and shaking a small head in puzzled wonderment. What a fool! Always working! Even in winter, the proper vacation time.

One-grooved *Pappogeomys* is larger on the average than most other gophers and is usually a dull yellowish brown that lightens on his face and underparts, giving him his common name, the yellow (or yellow-faced) pocket gopher. He has a quite broad skull that perhaps helps him in his bulldozing — at least he has a habit of heaping up larger mounds than most others.

He owes most of his allegiance to Mexico, eleven of his twelve species living in small scattered areas quite a distance south of the border. The twelfth, more ambitious, extends his range into big adjoining chunks of New Mexico and Texas and on up to take a piece of the Oklahoma panhandle and a corner of Colorado.

Two-grooved *Geomys* is usually referred to as the eastern pocket gopher because four of his seven species live in the deep southeast of the United States. Another reason could be that his major range where he lives in greatest numbers is the east of the American West: that vast region of prairies and plains extending from the Mississippi to the Rockies and from Canada all the way down to the lower Gulf Coast of Texas. Wherever he lives, he is a deep-soil digger, one of the ablest of the gopher earth-movers, so prodigal of energy that he digs extra-wide tunnels, often keeps five hundred feet of burrow in active use, and is expert at annoying highway engineers by undermining pavements. All he wants to do is to get from one side of a road to another — but he does it in the gopher way, underground.

Nobody loves the gopher. Most nonhuman predators approve of him — as an edible item. We human predators accord him equal attention — but for the opposite reason. We disapprove of him. Campaigns against him by individuals and farm groups and rancher associations and irrigation district officials and highway crews and assorted government agencies have been waged for a century or more. At times in some states county commissioners have organized local boys into competing teams, offering a bounty per gopher tail and a dinner for the winning team. At one time the Federal Department of Agriculture became so desperate on the subject of gopher control that it seriously considered an expedient that had proved to be disastrous when tried as a means of pest control elsewhere — that of introducing the mongoose as a recruit in the predator ranks. Fortunately better counsel prevailed.

All the usual poisons and dozen of special gopher poisons have been developed. Bombs to be put in his burrows and traps from simple to Rube Goldberg style have been devised for his personal disbenefit. I heard recently of a single highway district board that used more than a ton of poisoned grain in one month trying to discourage his activities. Pumps have been designed and manufactured for the sole purpose of forcing cyanide gas into his tunnels. He probably has been and is the target of more means and devices aimed at his destruction than any other four-legged creature with the effrontery to interfere in human affairs.

The original American humans, the Amerindians, wiser than the rest of us in such matters, seem to have had some affection for him. At least he figures (and favorably) in a few of their onetime tribal tales. There is a Jicarilla Apache fable out here in my part of the country, for example, which suggests that on occasion he could be a friendly fellow. Lion was angry at Coyote, hunting him to administer proper punishment. Coyote tried this and tried that, hid in a bush, hid in a hole, was still afraid. So Gopher hid him in one of his pockets. Lion became suspicious of the bulge in Gopher's cheek. But Gopher convinced Lion that he merely had a bad toothache.

Again, from an Iroquoian myth: When the world had been destroyed by fire and flood, the Mole and the Gopher had survived by digging into the ground, so the Sky-Father summoned them to bring up new earth from the underworld to replenish the soil for the restoration of plant and animal life.

That Lion fable may be worth a chuckle, but the Iroquoian myth is worth much much more. There, wrapped in mythic symbolism, is a significant scientific truth. There, read rightly, is the reason we Americans, though we cannot achieve love for him, should respect and be deeply grateful to that mean-tempered solitary obsessive little worker, the pocket gopher.

Every now and then a naturalist lets creep into a discussion of the gopher a hint that his activities should not be completely condemned. Mention may be made that his burrowing (at least where it does not interfere with human affairs) is good for the soil. It increases water absorption. It loosens hard-packed ground. It mixes minerals and humus into the soil, increases fertility by addition of buried plants and animal matter, a process helped along by his habit of storing underground more of his food than he ever eats — or, if a predator nabs him, has a chance to eat. As his tunnels cave in and his mounds weather out the result is much the same as that achieved by a gardener with a spade or a farmer with a plow. I have found comments that the work done is a "fair exchange" for the plants destroyed, that it is "distinctly beneficial to soil formation and vegetational productivity," that the "thrift of wild vegetation" is dependent upon his "continued activity."

It is even being learned that the rancher who finds his grass dwindling and gophers increasing in number on his range and puts the blame on them is all wrong. He is thinking backwards. He should regard the gopher increase as a warning, an indication that he has been indulging in overgrazing. The cycle involved is simple. When grassland is overgrazed, weeds and forbs begin to multiply — and so does the gopher, who prefers them to mere grass. If grazing stock is kept off that range for a while, the multiplying gopher will do the weeding and the grasses will begin to come back.

His burrowing is good for the soil. So too, of course, is that of other small-size burrowers. But they do their digging primarily to have homes, while he devotes almost his entire life to the work. Every individual gopher does it, not family groups sharing quarters and territories. In fact, he does much of the burrowing for other burrowers who take over and use his abandoned tunnels. He is the boss American burrower, the chief of all our "natural cultivators."

Start with what one gopher does. (Averages again.) The average gopher mound is eighteen by twenty-four inches and five inches high and weighs twelve pounds — and three or four of them may be bulldozed into being in a few hours. Mounds alone may be no indication of all the work done; he may be shoving some of the dirt loosened and moved into a tunnel to be abandoned. I know of one undersized female (she weighed barely six ounces) who was bothered by poor underground drainage and built a mound to hold her winter quarters under the snow six feet long, four and a half wide, two high, containing eight hundred pounds of dirt. That was for her home rooms — and she had her underground foraging to do.

Think next of a localized population, the solitary individuals all busily burrowing, avoiding each other by their sensitivity to ground vibrations. In good gopher country, say a mountain meadow or a piece of virgin plain, the five or ten gophers per acre apt to be there will have "plowed" and to some extent "harrowed" and even "fertilized" the entire area to an average depth of one foot in just about two years. And the gopher's way as an agriculturist is not our way.

The gopher does not remove the native vegetation, baring the area, to substitute a one-crop planting that will be swept away at harvest time. He co-exists with the native vegetation, lives in a reasonable balance with it. His seemingly haphazard tunneling and mound-making leaves scope for the native plants to keep on seeding and rooting themselves. At the end of those two years that mountain meadow or piece of plain will have about as much — and much healthier — vegetation as it had at the start.

Here in New Mexico I am not alone in observing that the native bunch grasses grow better on the weathering remains of gopher mounds than they do elsewhere.

Consider now how many gophers might be at work at any time around the clock and around the calendar.

Population estimates are few and usually localized. Ernest Thompson Seton remains the only naturalist to my knowledge who ever made a real effort to assess the overall gopher population. Using only low figures for extent of range and for population density, then pruning these some more, he arrived at a possible gopher population of one billion. That was back in the first years of this century. With our human campaigns against him offset to a considerable extent by our other campaigns against his natural predators, he may be holding his own. Some naturalists nowadays put it that the best we have been able to do is "stay even" with him.

One billion little bulldozers moving earth — and being good for the soil.

Toss in now what I am assured is a fact: that, contrary to what Darwin asserted, the earthworm is not worldwide in native distribution, is not native to America, certainly not to western America and thus to the bulk of the gopher's range. True enough, there are some earthworms in my New Mexico garden, but like all now found in scattered places through the western states they are descendants of those introduced, sometimes deliberately, sometimes accidentally, by human immigrants. Yet the earthworm is regarded as the prime "natural cultivator," the major creator of vegetable mould, the myriad engineer who has labored through the ages working and reworking soils for the ultimate benefit of man. Who did the job here in western North America? I submit that the gopher shouldered most of it — and was willing to do it in regions the earthworm would shun and where the work was difficult and really needed.

Forget any attempts at precise figures. Think in large aggregate terms. The gopher, just about as he is now, has been a native American since the early Miocene. Through the millions upon millions of years since then he has been at work. No doubt his range has altered again and again and his total population swung through drastic cycles with the long slow shifts of geologic and climatic time. But he has been here working in myriad numbers wherever his kind of work could be done. He has seen inland seas form and age and dwindle away, mountain ranges wear down and rise again, and always, when and where in western America there was ground to be worked he has worked it.

When the Rockies rose and the detritus of weathering rolled out toward what is now the Mississippi River to form our plains and prairies, he was on hand to help turn raw earth into good soil, to enrich it, to cultivate it, to pay for his food by keeping the surface levels fertile and manageable for the plant life developing there. He was the pasture-keeper for the bison and the pronghorn, who roamed in their own millions, the meadow-maker and maintainer for the deer and the elk of the high places.

You might say that he was doing his full share to prepare vast expanses of this continent for exploitation by us modern featherless bipeds. I prefer: he was answering the summons of the Sky-Father to make this American earth more habitable for all forms of life.

LK·POWELL

Order: *Rodentia*
Family: *Erethizontidae*

A Prickly Pacifist

SOMETIMES I play the iffy game. For example, I say to myself: IF
uncle Josiah Wedgewood had not spoken up for him, nephew
Charles Darwin would not have been aboard H.M.S. *Beagle* when it
raised anchor in December of 1831 for a two-year that became a
five-year voyage around the world. Again and following: IF the
Beagle had not tarried for most of the five years along the coasts of
South America and the nearby Galapagos Islands, Charles Darwin
would not now be known as the father of the theory of evolution.

The first "if" above rests on recorded fact. Young Darwin could
not have accepted the post of unpaid naturalist aboard the *Beagle*
without the consent and financial assistance of his father. The latter
thought the project impractical and refused approval, even wrote
out his refusal. But he left a loophole. "If" — again an if — "if you
can find any man of common-sense who advises you to go, I will give
my consent." Son Charles was starting out to try to find such a man
when uncle Josiah, equipped with unquestioned commonsense,
came forward on his own and clinched the consent.

The second "if" is more a matter of inference, a spinning of a
theory of my own. Charles Darwin was only twenty-four at the start
of the voyage, uncommitted as yet in any scientific direction, not
much of a student, at the most an amateur naturalist whose only real

scientific interest to date was geology. The *Beagle* took him to, and for a good four years kept him in, precisely that part of the world most likely to develop interest in the other natural sciences and to stimulate thoughts about the variety and the relationships and the possible origins of the living creatures of this world.

Suppose the *Beagle* had taken him coasting about Eurasia or North America or even Africa. The flora and the fauna would have been reasonably familiar to him out of his European background. Plenty of differences to be observed, yes, but more just variations on basic models with which he was already somewhat acquainted, not models new and surprising in themselves. He might easily have been more impressed by the geographical and topographical differences, sufficiently so to be confirmed thereafter in geology as his primary interest. But the *Beagle* took him to the one major continent where all kinds of differences and particularly those of living creatures were the most distinct, where a European could hardly help being impressed and intrigued by the strange and puzzling life forms nature had there achieved. The *Beagle* took Darwin to the site of a vast evolutionary experiment — and so doing made him an evolutionist.

It was something like seventy million years ago that the land connection between North and South America was submerged and for some sixty-five-plus million years following South America was isolated by ocean barriers. During all that time, while the other major continents could share to a considerable extent their evolutionary progress through exchange of living creatures, South America had to go it alone, working with the very early stocks already established there when the break came, obtaining only a very limited supply of new stocks arriving by the method known as island hopping. Undaunted by such restrictions, South America went ahead developing its own distinctive additions to the great web of life woven about this spaceship Earth.

Had I another lifetime to live, I would devote some of it to trying to find my way about among the strands of that web that are the South American fauna — as in recent years I have been stumbling about through those that are the North American fauna. But time hovers over me, relentless, and that is as it should be. Like all things living I too have a death-date due, must make way for others, just as in forms and numbers beyond reach of the mind through years the same other living creatures have had to make way that I for my brief span might become and be.

And so I am content, with regard to South American fauna in general and mammals in particular, to try to know reasonably well only those three mammals that the southern continent, since restoration of the land bridge, has sent via Central America and Mexico to share citizenship here with me in the United States. The armadillo is one and I have discussed him. The opossum is another and I have discussed her. Currently I am considering and contemplating of the third, the misnamed porcus-pinnae, the spined pig, the porcupine.

He is not a pig, not even remotely related to the pig. One glance at his teeth or a look at his toothmarks on a tree and the fact is instantly apparent that he is a rodent with those capable evergrowing rodent incisors. But he is unlike all our other North American rodents and related to them only very distantly out of the long ago past. He was originally and until relatively recently a South American rodent — and South American rodents are a special group unto themselves.

Way back in the very early island-hopping years when rodents were just beginning to be rodents, some very primitive ones reached the southern continent. They must have been few but they were enough. On through the ages they spread out to colonize virtually all of the continent and radiated out into a wide array of families and genera. Only three of those families are known to most of us northerners even by name: the porcupines, of course, because some of them have become North Americans; the chinchillas because of their superb fur; and the cavies because one of them has been domesticated and become a worldwide pet and laboratory animal under the silly name of guinea pig. But the capybaras, who happen to be the world's largest rodents, and the pacas and the agoutis, the viscachas and the hutias and the tucotucos, the bush rats and the rock rats and the spiny rats who are not rats at all — who but the experts among us know much more about all those than that perhaps they really do exist?

As a group they deserve their own suborder within the order Rodentia — and they almost have it. They comprise most of the suborder Hystricomorpha, the "porcupine-shaped." The few Old Worlders completing the suborder, the gundis and the cane and mole rats (also not rats) of Africa and the porcupines of Africa and Southern Asia, are included — though they differ from the South Americans in structural factors and in habits — because they have by convergent evolution followed somewhat similar lines of development.

As so often happens in taxonomy, logic of priority alone justifies that suborder label. It is based on "hystrix," the Greek-Latin word for "porcupine." So the South Americans, who are much in the majority and have much the more varied and distinctive life forms, are listed under a label fastened to them from across the oceans. However, since such matters trouble them not at all, I know they would not bother to back me in any complaint.

After all, it is an old story, the applying and misapplying of Old World labels to New World animals. Our dominant North American porcupine (along with his southern brothers) has suffered his share of mislabeling. In fact, when Europeans were first exploring the New World, they were still mixed up about the Old World porcupines. They considered them to be hedgehogs.

From time immemorial they had been acquainted with true hedgehogs, who include most of Europe in their rather comprehensive coverage of the Old World — but are not found in the New World. All hedgehogs are insectivores, relatives of the moles and the shrews. But they are prickly little beasts. And so, when other prickly beasts were being talked about in Europe and some imported from Africa — as Topsell put it, "brought up and down in Europe to be seen for mony" — they were thought to be a larger and exotic kind of hedgehog. Though willing to use the name porcuspine, Topsell made it plain he regarded the animal as "an Hedghog of the Mountain." For nearly two centuries thereafter porcupines were usually listed, as did Goldsmith following Buffon, as "of the Hedgehog kind." While naturalists were noting more and more differences and expressing doubts, to the common people of Europe a porcupine, any porcupine anywhere, was automatically a hedgehog — a notion they brought with them to this country. Even today I find people who tell me that of course hedgehogs and porcupines are close relatives. And to this day various townships in the United States, particularly in New England, still have on their books ordinances offering bounties for "hedgehogs."

The scientific classifiers soon corrected that mistake for themselves, but for quite a while they kept all porcupines, Old World and New, in the one family, Hystricidae. They even insisted on keeping our particularly distinctive porcupine, our North American, in the Old World genus *Hystrix*. Audubon and Bachman were still doing that in the middle of the last century even though Baron Cuvier had ably argued decades earlier for a separate genus and supplied a label, *Erethizon*, which I translate very roughly as "one who swells in anger."

Eventually the classifiers realized they could not keep all porcupines in the one family. They divided it into two, properly retaining Hystricidae for the Old World porcupines, taking a cue from Cuvier and adopting Erethizontidae for the New World porcupines.

The current settlement, then, is that our major North American porcupine, a gift from South America, belongs to the New World family and has his own genus, *Erethizon*. What used to be regarded as various species of him have been lumped as subspecies into the one species, *E. dorsatum*, which name could be interpreted as meaning that his swelling-in-anger is located on his back — exactly where it is, plus sides and often tail, on all porcupines.

His range is extensive, covering most of North America north of Mexico. I find that one of the most interesting of the many interesting things about him.

When the land bridge was restored, he was one of the two kinds of porcupines who began to extend ranges into North America. Both of these porcupine pioneers at that time were tropical or subtropical animals. One was content to stop when he had colonized Central America and southernmost Mexico, to stay there and develop a mild variety of species. The other, *E. dorsatum,* kept on northward, ever northward, and nowadays is as much at home far north up around Hudson Bay and in the northern regions of Alaska as he once was far southward in the subtropics. As he progressed northward, he left behind him a curious gap in the overall porcupine range in North America. There are several species of porcupine in the southernmost part of the continent. There is *E. dorsatum* all through the main northern body of the continent. But through a very wide band across central and northern Mexico from coast to coast and on through our southeastern states from Texas to the Atlantic there are no porcupines. Between the ranges of the southern porcupines and of our northern porcupine there is a large no-porcupine's land.

Inevitably *E. dorsatum* had to cross that wide gap to get where he is today. Inevitably during the endless succession of generations and the slow expansion of range northward, he had to be a resident of that area for quite a long period. Yet sometime in the not-so-long-ago past he canceled all holdings there and kept on pioneering northward. I can think of only one explanation for such behavior. I think we can thank the glaciers of the ice ages of the last million or so years for encouraging him to become a permanent citizen of the United States and Canada.

Obviously he was the more adventurous and adaptable of the pioneers. Suppose that he had become established well up in what is now Mexico, when the great glaciers of one of the ice ages came creeping down out of the northland creating climatic changes far in advance of themselves. We know that at one time and another they did come deep down into what is now the United States and that at such times conditions here in the Southwest, for example, were much different from what they are now, particularly in terms of climate and vegetation.

Our *E. dorsatum* would not have been one to go into retreat, to let himself be pushed back southward. Stubbornness is one of his most persistent traits. Then too he might have found the changes gradually taking place to his liking. It must have been during one or more of the long-drawn glacial advance periods that he learned to live, made the relatively slight adaptations needed for him to live, in what I would call the climate-shadow of the great glaciers. Then, as they in turn and in time began their slow withdrawal northward, he must have followed them, lured on by the conditions and food he had learned to like, abandoning that wide gap across the continent that, changing too as the distant glaciers became ever more distant, no longer satisfied him.

Thus much is certain: he who was once and in evolutionary terms not very long ago a warm- even hot-climate animal nowadays insists upon a coolish climate with winters that are real winters. That is why he lives in most of Alaska and Canada. That is why he lives in the northern half of the United States. That is why he lingers further south only where high enough elevations provide the proper conditions — in the east the Appalachian highlands, in the west the regions dominated by the Rockies and the Sierra Nevadas.

Since all of New Mexico is high land, the lowest point in the state still half a mile above sea level, and since the Rockies in their southern outposts march down through the state, he lives here along with me — and as if to give tangible evidence has on occasion left quills in the nose and mouth of a family dog.

The Old World porcupines are mainly terrestrial, living out their lives on the ground. The New World are mainly arboreal, spending much of their time aloft in trees. The Old World porcupines have retained the original mammalian five digits on all four feet. The southerners among the New World porcupines, presumably for aid in the arboreal life, have broadened the soles of their feet and reduced the digits on each foot to a functional four equipped with capable climbing claws. The onetime fifth digit on each hindfoot has been replaced by a flexible pad that may be of use in their style of climbing. Our northern porcupine seems to be something of a maverick. He too has broadened the soles of his feet and reduced the number of digits on his forefeet — but he has retained the full five on his hindfeet. And, curiously enough, he looks more like some of his far-distant and remote relatives, the African porcupines, than he does like his southern cousins.

What even more obviously sets him apart from those southern cousins is his tail. They have longish tails, some of them prehensile tails. His tail is short and thick and muscular. Apparently they have preferred to keep theirs longer and more flexible for more agile tree climbing. His helps him some in that respect, serving as a brace for his hindquarters as he maneuvers in a tree, but he seems to have decided to adapt it primarily as a weapon, a warclub well spiked with quills. His immediate reaction when threatened is to swing his rear toward the enemy and attempt to strike with that tail. He backs into battle. His tactics are almost unique among mammals, paralleled to some extent only by those of the skunk. He attacks by going backward. He retreats by going forward.

Though not in a class with the beaver and the capybara, as most rodents grow, he is among the larger. With his extensive range and seven subspecies he can vary in size following the fairly common mammalian rule the further north members of species live the larger they are likely to be. He can vary from eighteen to thirty-two inches in length, with six to nine of those inches stout tail, and from ten to thirty pounds in weight.

His body is thickset with short legs, his head small, blunt, short-faced, with very small ears, and he appears to have almost no neck. His eyes are small and dull and he has hairy lips and long black whiskers. His rodential insignia, his incisors, are quite large and when he opens his mouth are rather startling because they are a deep yellow-orange. His front feet have long curved claws, longer than those of his hindfeet, and he toes them in like a bear.

He looks larger than he really is because he wears a bulgy coat. The first layer is of dense fine blackish brown underfur two to three inches thick. Overlaying this are coarse guard hairs up to six inches long. These can give him some mild variation in coloration. In my part of the country they give him a yellowish tinge. They usually look as if they have never been groomed, sticking out at all angles all over him.

On the ground he moves with a slow deliberate waddle — except when he is in a hurry (which is not often) and then he humps along as if he has never learned really to run. In tree-climbing too he is slow and deliberate; no free-swinging or jumping, no acrobatics, just careful movement from one safe position to another. His manners are atrocious. When eating he grunts like the pig he is not and slobbers and smacks his lips loudly. When drinking he pushes his mouth into the water like a horse and noisily gulps. Even when waddling along with no apparent cause for irritation, he often grumbles and grunts to himself as if he nourished a permanent grouch.

There is nothing dainty or graceful or handsome about him. He is a stupid, awkward, unkempt animal. He is also a fascinating and efficient fellow creature. Everything about him is adequate for his needs. With a single exception I will discuss later, he has solved the problems of existence with far less trouble and worry and work than most of the rest of us.

It was Thoreau, if I recall rightly, who said he made himself rich by reducing his wants, who sounded the slogan: simplify, simplify. *E. dorsatum* has had the same notions ever since he began being a porcupine.

He is a vegetarian, which means that his food is stationary, does not try to escape, does not have to be pursued and caught. That is true, of course, for all vegetarians, but he has reduced the vegetative diet to its simplest most constant most consistent form. He can digest just about anything in the vegetative line and thus can munch on whatever he finds tasty through the seasonal changes of the year. But these are only treats along the way and he is in no wise dependent on seasonal items. The staple of his diet is the inner bark of trees, which is available and easily located the year around. He particularly likes the inner bark of cool-to-cold climate trees: birch and beech and aspen and the evergreens. In the matter of food he is basically as well off in the midst of severe winter as at any other time of the year. No necessity drives him of bustling about to stock food in hideaways for what to other animals are the long hard months or of storing fat under his hide then making changes in his internal workings so he can hibernate through those months. He goes about his slow deliberate business in winter as in summer. All he needs is a good overcoat and he has provided himself with that. Caught aloft in a tree by a bitter blizzard that may last for days, even weeks, he simply stays there, snug inside his coat, food right at tooth, and lets the storm wear itself out.

The beaver has the same dietary advantage — and a fine fur coat too. But the beaver has to work like — well, like a beaver to get through the year. He has to cut down trees and haul them into place to build a dam to impound water and then build a house in that water and then cut down more trees and cut them up and haul them into the water to have a winter supply of food. After rigging such an artificial environment, he has to make frequent repairs to maintain it. From the porcupine's point of view the beaver is a fool. The beaver has simplified his diet and provided himself with a good overcoat. But he has made the mistake of depending upon something outside of himself for defense against predators: water, the beaver pond. Because he did not simplify an imperative need, that of protection, he has tied himself to prodigious labor.

The deer and his relatives share in the dietary advantage, though not exactly by preference. In winter they can at least get along nibbling on twigs and branches and tree bark. They have avoided the beaver's mistake, do not rely for protection on an artificial environment, thus eliminating that kind of labor. But they have made their own mistake, rely for safety primarily on acute senses and speed in flight. And that means they must be constantly on the alert and when pursued expend large amounts of energy — which is particularly precious in winter.

So runs the report down any list of the other mammalian vegetarians. They have the dietary advantage in their varying degrees, but the means they have for survival of predation in adequate numbers all involve — and usually in combination — such things as physical specialization, multiplication of progeny, much hard work, and much expenditure of energy. Virtually alone among them the porcupine has met the primary want, survival, by the single simplest method. One specialization suffices for him, eliminates any need for frequent and lengthy and worrisome family cares, for bothersome labors and frantic escapes. With that one specialization he has simplified existence, has made himself rich. He can afford to be stupid and awkward and lazy.

I am referring, of course, to his pinnae, his spines, his quills, what Dr. Bachman rather fulsomely described as "an impervious coat of mail bristling with bayonets." He has those bayonets, plenty of them, on his back and sides, on the top of his head and on his cheeks, and very definitely on his tail. Ernest Thompson Seton estimated that a full-weaponed adult has some thirty thousand — and with amusing mathematical logic figured that, since thirty are enough to defeat one dog, such a porcupine could defeat one thousand dogs.

Certainly he has for his size an impressive number of them and he maintains his supply by growing new ones to replace those lost. But they are not what gives him his shaggy unkempt appearance. When he is undisturbed they lie low in his fur, hidden by the long guard hairs. When he is startled or provoked, they are pulled erect by an intricate musculature close under his skin — the swelling-in-anger of his genus name. Even then they are not very noticeable because they are not quite as long as the guard hairs. But they are there, very much there, in defense position. Short-legged body close to the ground, head tucked down between his forelegs, quills erect, he is well defended all around with his best defenses at the rear, which he seeks to keep aimed at his attacker. If he has managed to head himself into some kind of a cul-de-sac, say between the protruding roots of a tree with only his rear and its fast-swinging war-club exposed, he is virtually impregnable.

Ah, yes, you might say, that is good, very good, very clever of him to have developed those quills. But has he not made a mistake in the way he wears them? Ought they not always be erect in defensive position? Since they are not, is he not exposed to a sudden swift surprise attack striking him while they still lie low in his fur?

I would like to see anyone who believes that try the trick, try to surprise him and grab him before his quills are ready for action. As he waddles about in his slow deliberate way, he may seem not to be paying much attention to what is happening around him — and he may well not be. His eyesight is quite poor anyway. But he has his own kind of alarm system. To go about with quills always raised would be both a nuisance and a waste of them. They would be catching on things, picking up things, getting knocked or pulled out. He has no desire to be an animated pincushion except when necessary. So he has made those long guard hairs, which are longer than the quills, highly sensitive to tactile stimuli, responsive to the slightest touch. That is why he wears them sticking out all over him, giving him that unkempt look. They create a sort of protective aura around him. He knows what they should be feeling — or not feeling — as he waddles about. At the barest indication of a wrong touch, by muscular reflex in a flicker of a fraction of a second his quills are rising in anger. Suppose a predator slips up undetected by the porcupine's poor eyes or better nose or even better ears and tries a swift pounce. By time that predator's claws or jaws are trying to close on the intended victim, the porcupine's quills are rising erect to meet them.

Each quill has three parts: a solid pointed tip, usually black; a hollow shaft, invariably white; a root like those of a fur-hair of which it is an adaptation. The tip itself is a diabolical invention. It is needle-sharp and smooth for the first small fraction of an inch, then covered around with hundreds of tiny barbs aimed downward towards the shaft. Normally these barbs lie close pressed down, but when the tip penetrates into anything a backward pull makes them catch and anchor the quill. In fact, when they encounter warmth and moisture, say in the flesh of a predator, they flare out of their own accord for just such anchoring. Just below the barbs the tip swells a bit to join the shaft and form that part of the whole which many Indians once used in decorative craft work.

Since the shaft is not as firmly fastened at the root in the porcupine himself as the needle-sharp tip is the instant it is imbedded in anything, the quill is promptly pulled loose and remains where the tip has implanted it. If that anything is a living creature, that creature has suddenly acquired trouble — and trouble continuing long past the pain of impact. Even if the quill can be reached, taken hold of, pulled out, because of the barbs the pulling out is very painful and enlarges the wound. If it cannot be reached, the trouble intensifies. Again because of the barbs there is only one way the quill can move of itself and that is forward. With each movement of the flesh in which it is imbedded, it creeps forward, penetrating deeper and deeper, traveling on the average of an inch every twenty-four hours. Eventually, if the new unfortunate possessor of it is lucky, it will work its way out on the opposite of whatever part of him it entered. But if on its way it encounters a vital organ, he may die. And if it and others are clustered in and about his mouth, he may be unable to eat and confronts the ordeal of starvation.

Deadly weapons, those quills. For a not-very-big animal who has no need to kill for food and who, as a vegetarian, is a low man on the food-chain totem pole they come close to being the perfect weapons. They reduce self-protection to the simplest and among the most efficient terms among mammals. Other animals using other means of defense have to use these frequently. Merely by possessing his quills the porcupine seldom has to use them. They have given him a reputation among his wild fellows, including even the large predators, that in itself helps protect him. Almost all of them learned long ago that an attack on him has a high price and normally they pass him by. Only in hunger-driven desperation do they try and sometimes succeed in making a meal of him — and of those who do succeed some pay the final delayed penalty. There are records of coyotes and bobcats and wolves and lynxes and even pumas found dead, slain by the slow deadliness of porcupine quills.

Apparently of all American carnivores only the fisher, that sleek lightning-fast northwoods big cousin of the weasel, dines on porcupine meat fairly consistently and with fairly consistent immunity. The fisher has learned to slip a swift paw under and flip a porcupine over on his back and attack the soft unprotected underside before his clumsy victim can regain fighting position. Even so, the wise fisher usually takes the precaution of trying to provoke his prospective dinner into becoming careless from anger and frustration and exhaustion before closing in for the kill. I have heard tales that a fisher catching a porcupine on the ground in deep snow may burrow through the soft stuff to attack from underneath. It is said too that the fisher has developed an ability to pass through his digestive system any quills he may swallow while eating the flesh. That is probably true: I have read reports of quills compressed into small bundles presumably for being moved along found in the stomachs and lower intestines of fishers. I have also read reports of other fishers who acquired quills in other portions of their anatomies and paid the full final price for their meals.

All vegetarian mammals in the wild are subject to predation, but I believe the statement is sound that here in North America *E. dorsatum* suffers the least from it. His mortality rate through his active life is low. He has a good chance of living to a ripe old age — even of achieving that rare goal, natural death from old age. And along the way he has a remarkably easy leisurely time of it. By simplifying his means of protection and matching this with a simplification of diet, he has eliminated much worry and work for himself.

His mate has done some simplifying of her own in regard to maternal duties. To maintain an adequate porcupine population she needs to produce no more than one infant a year, a porcupette. Moreover she has so arranged her pregnancy that her infant is born almost ready to take care of itself. Its eyes are open, teeth are pushing through its gums, and it is already equipped with quills that are still soft to ease the birth but that in a few hours will be stiffening into service. Within a day the porcupette will be climbing trees after its mother and within a few more will be weaning itself by sampling vegetative tidbits. In not much more than a week it will be capable of becoming an independent member of the porcupine tribe. It will still follow its mother about for a while, but is no longer dependent on her for protection or for food.

Ingenious weapons, those quills. In their various versions they are the distinguishing feature of all porcupines, Old World and New. Like most unusual animal specialties, they have been the basis for curious beliefs.

The oldest and best known was and is (some folk still cherish it) fairly understandable. Though the atom bomb has begun to alter our way of thinking on the subject of armaments, it has always been difficult for us humans to conceive of any creature having weapons that he does not actively use — weapons which just by being he hopes will serve as protection enough. Obviously, we tend to conclude, he must do something with them. Add to that kind of thinking the remarkable ease with which anyone tangling with a porcupine can find quills stuck in some part of him without being sure just how they got there and it is all but inevitable the illusion should have arisen that the porcupine does do something with his quills, that these are not so much bayonets as they are arrows — that their owner can "shoot" them, send them flying some appreciable distance at and into an enemy.

Aristotle was, I believe, the first to put that down as fact. Later writers on through the Middle Ages repeated it with increasing exaggeration. In Topsell's time it was still as firmly imbedded in what passed for natural history as ever a quill was imbedded in flesh. Topsell's porcuspine had "quills or thorns upon his back which he shooteth off at his pleasure." The notion was not confined to Old Worlders thinking about Old World porcupines. It had been circulating independently among some Indian tribes of the New World about New World porcupines — in particular about E. *dorsatum.*

Goldsmith, by his time, knew better. He noted that "of all the porcupines that have been brought into Europe, not one was ever seen to launch their quills, and yet the irritations they received were sufficient to have provoked their utmost indignation." He concluded that the belief they could "discharge" their quills at will "has been entirely discredited of late." And then, having demolished that belief, he proceeded to cite as true two others just as deserving of demolition. One was that the porcupine "moults" his quills "as birds do their feathers" — a procedure which E. *dorsatum,* for instance, would never be fool enough to adopt since it would leave him periodically with armament rather seriously depleted. The other has its own morbid fascination.

> Travelers universally assure us that, between the serpent and the porcupine, there exists an irreconcilable enmity, and that they never meet without a mortal engagement. The porcupine, on these occasions, is said to roll itself upon the serpent, and thus destroy and devour it.

I suspect that the travelers supplying such assurances were reporting what they thought ought to happen, not what they had ever witnessed. They probably had in mind the "porcus" part of the porcupine's name and their conviction he was a kind of hedgehog. Wild pigs devour serpents and hedgehogs will do the same to those small enough to be caught and conquered. Why not then assume that a large kind of hedgehog somewhat superficially resembling a pig should do the same. And what more seeming logical than also to assume that a creature carrying his weapons on his back would roll over on a lowdown groundhugging opponent to bring those weapons into play.

There is not even seeming logic, however, in the basic assumption involved: that there is an "irreconcilable enmity" between the porcupine and the serpent. That must have been a hangover from both the medieval bestiaries and the writings of the early naturalists of more modern times, which were liberally salted with references to implacable animosities between various creatures supposedly determined to annihilate each other on sight. What intrigues me, however, is that Goldsmith should so calmly and with never a demurrer have been willing to attribute to the porcupine that kind of single-minded ferocity. In the long preceding paragraph he had just been discussing the porcupine in quite different terms — giving him credit as one who uses his "formidable apparatus of arms" only to defend himself.

On that point (no pun) Goldsmith was right and all we fellow creatures should be grateful that porcupines make him right. If our *E. dorsatum,* for example, who is one of the best armed of them all, were fast-moving and skittish and rambled about a great deal and was right ready to wade into a fight on slight provocation, he would be both an infernal nuisance and a positive danger. Instead he is obligingly slow and sluggish, not much of a traveler, and he never starts a fight except perhaps with another porcupine in the mating season — and then he merely sits up on his haunches with tail for a brace and boxes with his opponent without bringing his real weapons into action.

True enough, when he meets someone else as he waddles along, he does have a sort of irritable arrogance that is really a dim-witted complacency about his weaponry, which makes him take for granted that someone else should get out of his way. But what does he do? He stops. He raises his quills. He stomps his front feet. He shakes himself, rattling his quills. He may seem to be threatening dire actions. He is simply asserting identity and trying to make the situation plain. In effect he is saying: "Look, whoever you are, I am that prickly package that is dangerous only if you come too close. If you don't bother me, I won't bother you. And now, since you are undoubtedly more nimble and more energy-wasteful than I am, you be the one to step aside."

If such a threat or appeal or whatever you wish to call it fails, he may even turn about and make a few tentative advances-in-reverse, twisting his lowered head to peer back and see what effect these have. But if the someone else has the temerity to continue the dispute and seems disposed to attack, *E. dorsatum* will be the one to make way, to look for a good place to take a defensive stand.

I first heard the perennial porcupine joke more than half a century ago when I was young enough to be proud of myself for understanding what was supposed to make it humorous. First came the question: How do porcupines make love? Then the answer: With difficulty!

By the time many years later I had had the same weary witticism inflicted on me by some self-panicking would-be witster for about the tenth time, I realized some people thought it was humorous precisely because they thought it was true. They conjured up in their minds visions of a pair of love-struck porcupines perpetrating weird contortions and strange acrobatics trying to avoid each other's quills and achieve copulation.

Nonsense, of course. The female keeps her quills flattened down in her soft underfur and raises her tail up over her rump so that its quillless underside is uppermost. The male mounts her in what is known as the classic quadrupedal position — with this difference: he does not lean far forward as do most mammals, clutching her with his forelegs or biting or taking a grip on her neck. He remains more upright, crouched down on his hindlegs braced by his stout tail. Obviously he is now in a vulnerable position, the unprotected portion of his anatomy exposed to her quills should she choose to use them. She has no intention of using them; she is as eager as he is, often more so, and is concentrating on assisting him by pushing back against his forward thrusts.

Their difficulty is not in making love. Their difficulty is in *not* making love — in waiting for the right time for it. I suspect that is the worst worry of their otherwise much simplified lives.

There is only a brief period, a matter of only a few hours of only one day, during which the female can conceive. But there is a long period of preparation before, a sort of warming-up time extending into months, and another longish period after, a sort of cooling-down time. That makes the process difficult for them because they are strict believers in sexual intercourse for procreation only.

Early summer is her carefree time when she has practically no worries at all. Her this-year's porcupette is no longer tagging her. As always there is food aplenty and right now seasonable tidbits are available. In her cumbersome way she gambols about, plays games for herself, rolls on her back and bounces a stick with her feet, stands on her hindlegs and performs an awkward dance, beats out a rhythm with her forepaws on anything that will make a satisfying noise — in general indicates that she finds life well worth the living. Meanwhile, since porcupines are mildly sociable, occasionally even sharing sleeping places, she gets to know others in the neighborhood — among them the one of the males with whom she finally will mate. That seems to be more important to him than to her. He has what seems to be an inhibition that prevents him from mating with a female with whom he is not well acquainted.

Her carefree time, fine while it lasts, is limited. She will not mate until rather late in the fall, but by midsummer the process leading to her brief conception period is beginning within her. As it progresses she becomes increasingly nervous and irritable, a new urgency shows in her actions, she does seemingly foolish and erratic things for no other reason than to try to ease the restlessness permeating her. When summer has slipped into early fall, she knows what is bothering her; the urgency is becoming localized in her genitalia. Yet the right time, the only time for conception, is still weeks away. She seeks relief by pressing herself against anything available, a rock, a stump, a hump of ground. She takes up a fallen branch, stands upright on her hind legs with the upper end held in her forepaws, the other end tucked between her hind legs and dragging on the ground, and moves about riding it like a witch on a broomstick, rubbing her genitalia against it. As time passes she seems to be almost in torment.

All this while the male has been increasingly aware of what is happening to her. Like most mammalian males he is programmed for his sexual response to be triggered by her — and she has been triggering it. Restlessness and urgency have been rising in him too. He has been making tentative advances toward her and has been rejected. By time she is leaving her estrus scent about he too is in a kind of torment. Sniffing, he inspects everything she has rubbed against, even picking up pieces of dirt in his forepaws to hold them to his nose. He too seeks relief by riding a stick. Just as she is, he too is being frustrated by the inflexible rule that intercourse is for procreation only.

The right time is approaching. She has chosen him and stays near him. She abandons almost all activity, even eating, seems sad and withdrawn within herself. Presumably she is mobilizing her female resources for the all-too-brief period of release.

He is guessing now whether the time is right. He tries an experiment. He sniffs all around her, rubs noses with her. He stands on his hindlegs, penis erect, and moves toward her. He has guessed wrong and she shies away. More tormented waiting and he tries again. The time is closer now and she too rises on her hindlegs and faces him. Only a short distance from her, upright, penis erect, he ejaculates fine sprays of urine toward her. But the time, though close, has not yet quite come. She objects to the shower, grunting in exasperation and turning away. More waiting and he repeats the maneuver. Perhaps this time, perhaps the next, he has guessed right. She has only perfunctory objections to the shower. She moves toward him and he toward her. Still standing upright, gently they touch noses — and as if that touch confirmed all preparations, released all inhibitions, they sweep into action. With speed startling in such normally sluggish animals, she is down and in position and he is mounting her.

They will probably repeat their coupling several times or more during the next few hours but their day of dalliance is brief. Five hours is the upper limit of her receptiveness, less for some females. Her internal clock strikes. If she is going to conceive this season, it has happened by now and there will not be another chance until next year. She rejects him. Experiments and maneuvers on his part are useless and he knows it. Romance has come and gone.

But the aftereffects remain. Both of them now go through their cooling-down periods, reversals of the warming-up though not as long-drawn, the urgency diminishing, the restlessness and extra irritability gradually fading away.

I wonder whether Mr. and Ms. *E. dorsatum* ever wonder whether their brief annual sex spree is worth the bother and frustration of before and after. Perhaps she feels a bit the better about it because she at least has the presumable satisfaction of producing a porcupette in whom for a while she can have some pride. Her temporary spouse has none of that and very likely does not recognize his progeny if they happen to meet.

Certainly the porcupine version of the mating process seems unnecessarily annoying and complicated when compared with those of most other mammals, especially most other rodents. It seems less an asset for survival than a taking of chances with survival. Could it be considered an evolutionary mistake? Perhaps not so much a mistake as a side-slip, a tangling of adaptations which has not yet been straightened out. Perhaps this too is related to their relatively recent transformation from warm- to cold-climate creatures. Thus:

When they lived in a year-round warm climate, they had already simplified life for easy living. Armed with their "bristling bayonets" guaranteeing a low mortality rate, taking on family cares once a year with only one infant was enough. By developing a long (for a rodent) gestation period of four months to produce an infant already well along toward self-sufficiency, Ms. *E. dorsatum* had reduced maternal cares to a minimum. Neither she nor her spouse had any need to make burrows or nests for themselves and their progeny. Birth of that progeny could be any time of the year. Suppose, as seems possible, even probable, they had become adapted to having the annual sex-urge rise in the midyear months, a pleasant period for romance, and in the usual manner come fairly quickly to its copulation climax. Then they moved northward and confronted increasingly a definite problem — winters.

No parents with a modicum of regard for the welfare of their progeny would deliberately schedule birth of the same for the toughest time of the year. Dependable old natural selection, basic assumption of orthodox evolutionists, would apply its pressures. If the *E. dorsatum*s gave way to the urgings of their sexual upswing previously programmed for midyear, their porcupette would be born just when winter was clinching its hold in the regions they were now pioneering. What to do? Various other mammals have solved the problem of separating copulation and the start of gestation by their females' developing either delayed fertilization of the ova or delayed implantation of the fertilized ova in the wombs. Such methods enable them to have their romancing when their pro-gramming commands, then to postpone the start of actual gestation. The *E. dorsatum*s have either been too stupidly backward or have not yet had the evolutionary time to adopt either method. With the an-nual sex-urge still taking them under the old timing, they have sub-jected themselves to the frustrating method of prolonged postpone-ment of copulation. And perhaps because the pressure within them has been pent so long, their time for release of it is so explosively short.

I am told now and again that porcupines are irritable and rather solitary because they have quills, that a body covered with prickles just naturally engenders an unsociable disposition. Nonsense. The most unsociable of rodents are the gopher and the kangaroo rat and mouse and they have nary a prickle. I suggest (as psychologists often suggest in regard to humans) that the porcupines' self-im-posed sexual discipline is responsible for their touchy dispositions. I think they have a right to be irritable and short-tempered much of the year — to be what Shakespeare called the Old World version, the "fretful porpentine."

On the average *E. dorsatum* has had fair luck in his relations with us humans. His fur has no commercial value and he is not a trapline pilferer, so trappers have ignored him. His flesh, though formerly on the menus of some Indian tribes, has never been highly re-garded, so hunters for food have not done much to boost his mortal-ity rate. Being slow and stupid, he offers no challenge to the person with a gun, so hunters for sport have never considered him a game animal — and since he is not a competing predator, they have never clamored for porcupine-control projects. I suspect that throughout his northwoods career his major nemesis has been forest fires.

Even when condemned for some of his habits he has had his de-fenders.

For example: he has frequently been charged with damaging forests, ironically enough by those who have damaged them infi-nitely more — lumbermen, timber merchants. Dining on inner bark, he can cripple trees. That usually is not very serious because he seems to know that he can make trees work for him. When he has gnawed away a patch of bark creating what is known as a "cat-face," he usually moves on in the seeming knowledge that the scar left there will block the downward flow of sugar through the inner bark and an extra quantity of it will collect just above the scar to provide him with a fine meal on a later return trip. But the fact stands that he does not only cripple but kills some trees and that annoys the lumbermen who want to do precisely the same thing in their way for their purposes.

The tree-killing charge is true enough, admit his defenders, but in an overall sense he is helping the forests. He is performing an ecological function, thinning out stands of timber so that other trees have a better chance at healthy growth. Moreover, as a tree-climbing vegetarian who operates on through the winter, he provides food for other vegetarians at the time they are hard-pressed. As he does his foraging aloft in his clumsy, sloppy way, he sends down a shower of twigs and small branches, especially from conifers with their ever-green needles, supplying browse for such groundbound fellow citizens as deer and hare and rabbits.

Again: he can certainly be a nuisance to people living or spending time in the woods, to rangers and hunters and campers and vacationers and such. Because of some nudging but obviously not too serious a lack in his ordinary diet he has an inordinate taste for anything salty, even for anything with just a tinge of saltiness imparted by nothing more than the touch of sweaty human hands. He will mess up an unoccupied cabin, gnawing a way in and going to work on whatever his sensitive nose says suggests salt. Being complacent about his quill-protection, he will invade a camp and, though chased away, may repeatedly return. He will chew vigorously on camp stools and tent poles. An ax handle will provide a snack. Anything leather delights him, particularly gloves. Having poor eyesight, content to pay attention only to moving objects, he has been known to start dining on a saddle while a sleeping camper is using it for a pillow, even to come up to some person who is standing still and start chomping on that person's feet-filled boots. And of course, if a dog happens to be about, trouble of another and more serious kind is apt to occur.

All that is true enough, say his defenders again, but the very characteristics that make him such a nuisance also give him a special value all his own. He is the one wild mammal who can be quite easily cornered and killed by a human who is lost and starving in the woods and is armed only with a hefty rock or a stout club.

When I was a boy in the Great Lakes region, the tradition was strong: *Never kill a porcupine unless absolutely necessary.* I can recall a Scoutmaster squatting by a campfire and saying: "If you must get lost in any woods, make sure it has a good population of porcupines."

Primary human pressure on *E. dorsatum* has been and is the indirect but pervasive one afflicting most wild creatures: destruction of habitat as people and their appurtenances monopolize ever more of the available environment. This presses particularly hard on him because he is not among those who can get along close to human settlement. He is too big, too slow and stupid, too prone to blunder into inflicting injury on domesticated animals and unwary humans. In the words of Buffon, the porcupine knows only too well "how to defend himself without fighting, and to wound without attacking." He is not a tolerable neighbor and he is too clumsy and conspicuous to maintain an elusive hideaway existence. It was inevitable that in the early days of settlement, particularly in the then-still-forested eastern part of the country, bounties would be placed on him, some of which though largely neglected still stand.

Under what I call natural conditions (that is, before we humans became serious interferences) his population over any reasonable period remained in balance, kept there by normal mortality and the attentions of the fisher. There must be some sort of compensation involved in the fact that just the one predator had learned to be rather consistent in dining on him. If the others had regularly practiced the same trick, he might not have survived. Then we humans became deadly predators of the fisher, whose fur, especially of the females, brought record prices. As the fisher population declined, became practically nil in whole regions, the porcupine population began to increase. Slowly, slowly, because of that slow reproductive process. For quite a while the effects were not noticed except by oc-

casional woodsmen. But eventually, as we humans gobbled up ever
more of the timber resources and the price of wood began to soar
and more porcupines were crippling and killing trees in what stands
remained, talk of bounties began again.

This reached a sort of climax in the 1950s. The state of Maine returned to the bounty method, offered fifty cents per porcupine
killed with evidence of the killing to be the four feet. Soon town
clerks were pleading for additional funds to meet the payments and
struggling with the problem of disposing of the feet in some safe
secret manner to avoid their being found and presented again.
Then over in Wisconsin, where bounties were also being discussed,
wiser counsel prevailed. The fisher, extinct in the state for some
years, was reintroduced from more fortunate states. After a while
talk of bounties faded away; the fisher was doing his duty as nature's
agent for porcupine control. And a lesson was being taught that I
hope will have much wider application than merely to the porcupine
problem. Nowadays, with trapping coming under strict rules in
state after state and the fisher protected in some of them and staging
a comeback on his own (he never becomes very numerous), the porcupine problem is chiefly only a localized affair here and there in
places where new settlement impinges on areas he has been occupying.

I note that as early as 1954 the U.S. Fish and Wildlife Service asserted that *E. dorsatum* is one of our most interesting wild citizens
whose survival should be encouraged. Quite right. He is unique
among us, the one rodent who has come out of the vast evolutionary
experiment of South America, just as we humans here have come
out of our own beginnings elsewhere, to establish himself as a citizen
of North America. He did that long before any of us did. He is a
fellow immigrant — and he has priority rights.

Though we latecomers have pushed him out of much of his original range and encroach increasingly on what he has left, he holds no
animosity toward us. In his slow bumbling way he is willing to be
friendly whenever we are too, as Audubon and Bachman discovered
when they kept one in a cage in Charleston and occasionally let him
loose to wander about the garden.

It had become very gentle, and evinced no spiteful propensities; when we called to it, holding in our hand a tempting sweet-potato or an apple, it would turn its head slowly towards us, and give us a mild and wistful look, and then with stately steps advance and take the fruit from our hands . . . If it found the door of our study open, it would march in, and gently approach us, rubbing its sides against our legs, and look at us as if supplicating for additional delicacies. We frequently plagued it in order to try its raising its bristles at us.

Its bristles. Its quills. Always these come to the fore in any discussion of the porcupine. They constitute his trademark, his heraldic coat of arms. Goodrich, whose attitude toward our fellow creatures was in some respects ahead of his time, paid him the proper tribute more than a century ago.

The hare has his speed, the squirrel activity, the marmot caution, the beaver ingenuity, the rat most of these qualities; the Porcupine destitute of all, has his quiver of arrows, which he shakes in the face of his foe, to frighten him if he is a coward, and to pierce him if he has the courage to make an attack . . . Without his quills, the Porcupine would seem to be a singularly unmeaning, uncouth, and helpless sot; with them, he has a position in history, and figures in literature as the emblem of human fretfulness and conceit.

Note that last: "the emblem of *human* fretfulness and conceit." Inasmuch as *E. dorsatum* has those characteristics, he shares them with us.

And the more I ponder those quills, the more I wonder about the meaning of them. Did he develop them to compensate for his being stupid and awkward and lazy or is he stupid and awkward and lazy because he has them? In his case was necessity the mother of invention or was invention the parent of neglect of other faculties? Once finding himself reasonably safe behind a barrier of bristles, did he stop developing in other directions? Having achieved an outstanding adaptation in an early burst of creativity, was his capacity for further achievement so exhausted he could do little more — or was the incentive to do more simply eliminated?

But why, having achieved through simplicity the kind of reverse wealth recommended by Thoreau, should he ever be thought to have had any obligation to do more, to go on inventing additional adaptations that quite probably would only have complicated his existence? Are notions of the value of "doing more," of "progressing," of adding complications to living, merely illusions fostered by our own quite lately invented adaptation, the human brain? Those quills prick me into posing such questions — to which I have no answers. What I do know is that in an objective view *Erethizon dorsatum* is as contemporary and as complete a strand in the current web of life as is *Homo sapiens* — and is a far less dangerous and destructive strand.

All the same, I am grateful that none of my early primate ancestors achieved as simple and effective an invention as his did. Evolutionary movement toward me might have stopped right there or at least have been considerably slowed and I would not be sitting here right now congratulating myself on my ability to consider and contemplate of my fellow creatures.

Order: *Rodentia*
Family: *Heteromyidae*

Our Different Mice

MY MIND boggles whenever I try to think of the order Rodentia as a whole. It is so big, so complicated, still so much a field for disagreement among the taxonomists themselves. I can take it only in small doses, one small group or family at a time.

However figured, the order covers the whole of the habitable land of the world with more than 360 genera and about 1,730 species — which is almost as many species as all the other mammalian orders combined. No doubt as time marches on the lumpers will do more of what they have been doing lately, will trim down that species total by reducing some of them to subspecies status on the basis of increased and more accurate data. Nonetheless the assertion will probably always be sound that at least two out of every five species of mammals are rodents. (Another one out of the five is a bat, leaving just two to be shared by all the rest of us.)

Within the order itself at least two out of every three species are true mice or rats. These are so dominant and so numerous and many of them have been so familiar to us humans for so long that almost automatically we assume that any small rodent is one or the other. Time and time again we have applied mouse or rat labels to other small rodents who actually are quite different in vital aspects — certainly as different, say, as cats are from dogs. Which is what we have done with the small and highly specialized family of rodents of whom I am currently considering and contemplating.

In March of 1812 the English naturalist John Vaughn Thompson read a paper before the Linnaean Society in London in which he described a small rodent native to the American tropics. Superficially it resembled a mouse but it had certain peculiarities. He concluded that, yes, it was a mouse but a new kind of mouse and since it had coarse spiny hairs perhaps it was a connecting link between the mouse and porcupine. He labeled it *Mus anomalus,* which I translate as "deviant mouse."

During the next few years more was learned about this little rodent's peculiarities and in 1817 a French naturalist named Desmarest decided it deserved its own genus status. In choosing a label he made no attempt at emphasizing any specific peculiarity, simply made it *Heteromys* from Greek roots adding up to "different mouse."

Meanwhile one of the major peculiarities of this *Heteromys* was becoming the basis for a common name. He had cheek pouches opening externally like small-scale versions of those of the pocket gopher. Inevitably he became the pocket mouse.

Along in the 1830s a German traveler, Maximilian, Prince of Wied-Neuwied, came to the United States and accompanied a fur-trading expedition up the Missouri River to study Indian tribes along the way. Like many a 19th century gentleman of independent means (Darwin, for example) he was what I wish I were, a competent amateur ethnologist, botanist, geologist, zoologist. When he published his findings in handsome German and French and English editions and in papers for scientific societies, these included material on many subjects other than Indian cultures — and on many creatures other than Indians.

One of the creatures he described was a small native rodent superficially resembling a mouse but with certain peculiarities similar to those of Desmarest's *Heteromys* far away in the American tropics. In his turn Maximilian decided that his different mouse deserved genus status and he gave it a label, *Perognathus,* also from Greek roots, which I translate very freely as "abnormal jaw." His choice, I believe, was based on the fact that his different mouse (like Desmarest's) had twenty teeth while all other known mice (and rats) did very well with sixteen.

Here, then, were two small native American rodents with cheek pouches and twenty teeth and inevitably called pocket mice. But they had their own differences. Maximilian's became known simply as the pocket mouse. Desmarest's with its coarse spiny hair became the spiny pocket mouse.

Meanwhile again another native rodent with the same style cheek pouches and dental equipment had been receiving more and more notice. In 1841 an American naturalist named Gray gave this one proper recognition and a genus label based on another peculiarity: long strong hind feet and legs adapted for bipedal saltatorial locomotion — which is a fancy way of saying that this rodent moved about by hopping on its two hind feet. Gray chose *Dipodomys*, again from the Greek, which I translate as "two-footed mouse." Almost inevitably, given that hind-legs-only gait, this one would have kangaroo in his common name. Being larger than *Heteromys* and *Perognathus*, almost inevitably again he would be called a rat not a mouse. Kangaroo rat he became.

About this time the suggestion was being offered that these different mice and rats were different enough to merit a family of their own. Some classifiers began labeling them the saccomyidae, the "pouched mice."

Then in 1891 another native rodent who had the proper peculiarities but had escaped previous notice because he had a sparse and restricted range was recognized and given a genus name within the family by the American zoologist C. H. Merriam. This newcomer presented a mild puzzle. He was small like mouse-size *Heteromys* and *Perognathus*, but he had the bipedal saltatorial specialty of rat-size *Dipodomys*. Merriam tagged him *Microdipodops*, which I translate roughly as "tiny two-foot." Continued puzzling about him is reflected in his two competing common names, which seem to be about evenly matched in usage: kangaroo mouse and dwarf kangaroo rat.

There were now four known and named genera of these native rodents superficially resembling mice and rats but possessing among them definite peculiarities. In 1902 Merriam added another and gave it the genus name *Liomys*, which strikes me as a curious choice because the Greek roots translate into "smooth mouse" and the little animal is a variant of the spiny pocket mouse. I can only suppose that Merriam was emphasizing that *Liomys*, though not wearing an exactly smooth coat, did have a smoother one than Desmarest's original *Heteromys*.

Five living genera. No more have been discovered and probably no more will be. Theirs is a small select family.

For some reason the family name Saccomyidae has not found favor. For quite a while now taxonomists have been agreed on a family name based on Desmarest's emphasis that these are "different mice" and have listed them as the Heteromyidae.

I dislike that label — just as I dislike being forced by their commonly accepted common names to refer to these fascinating fellow creatures as various mice and rats. Even when I refer to them collectively as the heteromyids, I am still calling them mice. I am sure they themselves do not care one whit how I refer to them. But they distinctly are not mice or rats. They are not even placed by the experts in the rodent suborder dominated by the mice and the rats. If they must be associated by name with any familiar and long-known other rodents, they should be called squirrels. Most experts put them in the suborder Sciuromorpha, the "squirrel-shaped."

Actually their closest relative is the pocket gopher. They share with him not only pockets but a surprising number of habits and characteristics, so much so that some taxonomists put their family and his side by side in an exclusive superfamily. But even that kinship is rather remote in an evolutionary sense. The misnamed heteromyids have been going their own way for a very long time.

All of them have long narrow hindfeet and hindlegs longer than the fore. Their tails are long and haired. They have the general rodential jaw arrangement enabling them to use their longish incisors for gnawing and cheek teeth for grinding. Their skulls are somewhat broad, chiefly because the bullae, the auditory chambers of the middle ears, are proportionately larger than those of other rodents — in fact, I believe, proportionately larger than those of any other mammals. Their cheek pouches are furlined like the pocket gopher's and they fill and empty them in the same manner. They live in complicated burrows and are almost entirely nocturnal in habits.

Like the gopher, perhaps even more so, they are inveterate food hoarders, usually storing more food in their burrows when it is available than they manage to eat. They have a strong regard for private property and zealously guard their larders. As a result they are, again like the gopher, definitely antisocial and except for brief sexual interludes lead solitary lives, regarding any intruder, especially of their own kind, as a probable pilferer to be driven away. As another result, though they can be remarkably attractive pets, care in keeping them is required. Two of them, regardless of sex, if put in too close quarters, will almost certainly fight — and the female is deadlier than the male. They will get along and if of opposite sexes finally breed in captivity only if their enclosure is large enough to simulate natural conditions.

They are, then, confirmed individual territorialists. Yet they also seem to have a communal sense of social responsibility. When conditions are unfavorable, they practice birth control, reduce the number and size of litters or even temporarily inhibit reproduction altogether. Current studies suggest that estrogenic substances in fresh green plants are what stimulate the reproductive urge in them. In times of drought, then, though they can survive all right on what dried plant food is available, they slow down or stop breeding until the right kind of food is again available. If population pressures become too strong despite such precaution, they will reduce the size of their individual home territories to accommodate larger numbers within the same overall area — and when the pressure is less, expand them again. Thus, though always ready to fight for property rights, they also seem to try to hold down the occasions and necessities for fighting.

They will also do what seems to me surprising when they are at home in their burrows and any intruders (human or otherwise) come snooping about. They are small and subject to frequent and serious predation, yet, presumably on the chance the intruders are interested in taking over abandoned burrows or raiding storerooms, they immediately announce that they, the owners, are in residence. The heteromyid "mice" do this by squealing, the "kangaroos" by thumping with their long hindlegs. They seem unmistakably to be signaling: Occupied. Go Away. No Trespassing.

The fossil record on rodents is still spotty and subject to dispute, primarily because the evidence accumulated is still scanty — which in turn is partially the result of the early paleontologists being chiefly interested in the larger animals who left more noticeable and more manageable remains. But the record as deciphered to date shows that the heteromyids swerved off on their own evolutionary path well over 30 million years ago here in North America and have remained loyal Americans ever since. Their present family range covers considerable territory, from southwestern Canada down through our western states, on through Mexico and Central America and into northern South America, which later continent they began to colonize soon after the connecting land bridge was re-established.

Their habitats include prairies and plains and deserts, high grasslands and brushy regions, dry open forests and humid tropical forests. Five genera seems a small number for such small animals with close-to-home habits living in such far-spread and diverse environments. Given the size of their range, theirs is among the genetically more conservative of rodent families. Actually, to my way of thinking, though there are five genera, there are only three life forms: the pocket mouse, the kangaroo mouse, and the kangaroo rat.

Dropping down the taxonomic scale, however, they have done quite well with speciation. Leaving out the South American species about which I know very little, I tally for North America seventy-one species and at least three hundred currently recognized subspecies.

The more I learn about them, the more I appreciate what a close-knit little family they are. All of them share in all of the major family characteristics. Almost anything that can be said about any one of them applies to all of the others in some degree varying from mild to extreme. As a family they seem to me to have been remarkably ingenious in developing techniques for survival under difficult and dangerous conditions.

Of the three genera of pocket mice the two spinies represent the southern branch of the family. Desmarest's *Heteromys* is strictly tropical, inhabiting southernmost Mexico and Central America, and is the one who has extended the family range into South America. Merriam's *Liomys* favors more diverse environments. While he is found as far south as Panama, he also has the rest of the family's general liking of dryer, semiarid and even arid areas and he is more numerous up through Mexico and has even crossed the Rio Grande into the southern tip of Texas. Apparently he is the most primitive of the heteromyids — that is, he is thought to be the closest to the family ancestors.

In the family scale he is medium-sized, averaging about seven to nine inches long with half of that a fairly well haired tail. His body coat is composed of what can be called normal fine hairs but these are almost hidden on his upper parts by other longer stiffish and flattened and grooved bristly hairs, which are not really spines but give that appearance. In coloration, tail included, he is greyish brown above, white below. His hindfeet and legs are longer than the fore, indicating that he made a start toward the saltatorial specialty other members of the family would carry to an extreme. When he hurries he hops — but he uses all four feet. He has a small specialty of his own, spoonlike claws on the second digits of his hindfeet which may be a help in digging but are probably more useful for combing his grooved "spines." Like all members of the family he indulges in frequent dustbaths followed by careful grooming.

By current reckoning *Heteromys* has 10 species, *Liomys* 11. The third genus of pocket mice, Maximilian's *Perognathus,* outdoes those two together with 26 species and more than matches them in extent of range. He is at home, preferably in arid and semiarid areas, from well down in Mexico northward through all of our western states (with possible exception of Montana) and on into southern Canada. On his way north he seems to represent a gradual shift away from some aspects of the basic *Liomys* model. His more southern species are only slightly smaller than those of *Liomys* and still have vestiges of spininess in the form of bristles on their rumps. His more northern species are still smaller and have foregone the bristle fashion altogether.

As the above suggests, *Perognathus* can vary quite a bit in size. His largest species can measure up to seven inches over all, his smallest about four inches. In all of his species he has gone further than both *Liomys* and *Heteromys* toward saltatorial specialization. His tail (important for balancing) is longer than theirs in proportion to head-and-body length as are his hindfeet and legs in proportion to the fore. He is a better jumper — but still uses all four feet for locomotion. When he is in a hurry, the differential between his two sets of legs gives him a jerky mechanical-seeming gait. When he is at work, it gives him an advantage. He can sit back on those long hindfeet braced by that longish tail with his forefeet free to scrabble in the dirt and pick up seeds to be tucked into his pockets.

I am tempted to say that the speed with which he can fill those pockets has to be seen to be believed. That fact is it cannot be seen. Those little forelegs ending in what for this kind of work are virtually hands move so fast that what is seen is simply a blur. He can fill his pockets even with tiny grass seeds in a matter of seconds — which means he does not have to be out in the open in any one spot exposed to predation for more than a very brief time. He can empty the pockets even faster with a few blurred sweeps of those forehands — which means that in an emergency he can get rid of the extra weight almost instantly and be stripped down for flight. But even when undisturbed he rarely slows down. He will fill his pockets, dash to his burrow, stow the cargo in a storeroom, and be back for another filling in less time than it is taking me to type this sentence. Checks made of some of even his smallest species have tallied almost three thousand seeds gathered and stored in an hour.

One of the heteromyid survival techniques *Perognathus* has made quite effective is an ability to slow down his metabolic processes when times are tough and it is advisable to conserve energy. In scientific papers he is said to be able to hibernate and to aestivate — though he does neither in the usual meaning of those terms to a layman like me. That is: he does not slow down his vital processes to "sleep" away either the winter or the summer. He slows them down for brief periods only, probably never more than a week at a time, and does so any season of the year there is reason for him to do so — during really cold cold-spells or when the ground is snow-covered, shutting him off from food gathering, or when food is temporarily too scarce to sustain his usual activities. Confronted with such an emergency, whatever the season, he reduces his energy expenditure to wait it out.

Even under good conditions his body temperature has a normal range of fluctuation of as much as 9 degrees Fahrenheit, rising when he is active, dropping when he is at rest. That can be called his first line of defense against energy depletion when an emergency arises and he has to sit it out in his burrow. The second is the food he has stowed in his storerooms. The third and probably final recourse is the hibernation-aestivation business. If the weather is warm and the emergency persists, he can lower his body temperature below its normal range, slow down his heart beat and respiration, and go into a kind of torpor with energy consumption reduced. If the weather is cold he can lower his body temperature even more and sink into a deeper torpor with energy consumption even more reduced. But in either case he does not have to go through the slow process of the necessary physiological changes common to those animals who go into true seasonal aestivation or hibernation. He can accomplish the change in a few hours — and come back to activity at about the same rate.

That much about his energy-conserving resources has been established by careful researchers. They have not yet, to my knowledge anyway, discovered how he knows when to bring himself back to normal activity — or, to be more accurate, what stimulates the return to activity for him. If he is waiting out a bad cold spell (or period of snow cover), I suppose a rise in the external temperature (he does not dig deep burrows) might trigger a response. But if it is summertime and he has resorted to his version of aestivation because the current food supply of his little territory is exhausted, what tells him when a new supply is available — when, for instance, another kind of grass or weed has come into seed? Does he simply hold out in his torpor until his own internal situation nudges him with the news that he is close to the end of his resources and it is now or never for a try at finding food? Whatever his signal system, it works well. He is a small animal under constant pressure from every kind of predator inhabiting the same regions. He lives where severe cold spells and equally severe hot spells often occur and in deserts where seed-producing plants are sparse and often have long waits between the rare rains to bring them to fruition. Yet he survives — and in surprising numbers.

Here in New Mexico we have members of eight of the *Perognathus* species — and at least eight subspecies unique to the state. Among them they show the whole extent of his size scale. My favorite is the species whose name, *flavus,* means "golden yellow," which he justifies by wearing a coat distinctly gold tinged on its upper parts. He also justifies his common name, silky pocket mouse, by composing that coat of fine soft hairs. He is the smallest of all the pocket mice, certainly the smallest of all southwestern rodents and possibly of all American rodents. He is right down there close to the shrew class. A healthy adult may weigh all of a third of an ounce. Though almost as pugnacious as his larger fellows toward his own kind (always possible pilferers), he has no animosity toward us humans. "Few small animals are more beautiful than these silky, bright-eyed mice," was Vernon Bailey's verdict. "They are timid and when caught in the hands will struggle to escape, but make no attempt to bite or scratch. If held gently they soon become quiet and may be stroked as they sit in the open hand."

Not long ago six members of another of *Perognathus*'s species won a special distinction for him. They became astronauts. They were Californians of the species *longimembris,* so named for especially long hindlegs but otherwise almost as small as *flavus.* The six were given passage aboard the Apollo 17 space-flight with tiny detection devices implanted under their scalps to test whether cosmic ray particles would be injurious in prolonged flights. Why were they chosen? Because, as I have been stating, they can withstand rigorous conditions. And because they are small and would add no appreciable weight. And because they need no water.

Need no water? That is a heteromyid survival technique best discussed in connection with another member of the family.

"There is also a different breed of mice," wrote Aristotle long ago, "that walk on their two hind-legs; their front legs are small and their hind-legs long." He could have been describing the "kangaroo" heteromyids of North America. Actually, of course, he was thinking of the specialized Old World mice known nowadays as the jerboas — in his case those of North Africa in general and probably Egypt in particular. Their similarity to our "kangaroos" is another case of convergent evolution taking place in widely separated parts of the world.

We have another example right here at home. One of our specialized kinds of true mice usually known as jumping mice have similar similarity and Aristotle could just as well have been describing them.

So our "kangaroo" heteromyids are not the sole rodential inventors of bipedal saltatorial locomotion. As a matter of fact, the springhaas of North Africa and the Near East, the elephant shrews of South Africa, and the jirds of Mongolia all also do quite well with it. But I believe that Gray's *Dipodomys*, our kangaroo rat, may well have done the best job of developing it for efficient action. That would be in keeping with his general nature. He carries most of the heteromyid habits and characteristics to the family extremes.

He is not really a big fellow, though he is the biggest of the family. His species will vary from about 9 to 14 inches in length — which is a somewhat deceitful statistic because so much of his length is tail. He has a short stocky body and a head almost as big as that body and a major reason for its bigness is that his bullae are proportionately very large. His tail is always longer than body and head combined, in some species as much as half again as long. It is well haired with a tuft on the tip and is usually dark above and below with white stripes running along the sides. The rest of him is species-varying shades of tan above shading to white below and sometimes he has a white band across his thighs. His eyes are large and dark, appearing almost pure black, and are set well back on the sides of his head. Usually there is a pair of dark spots on his forehead that can give the impression of a second set of eyes. His forelegs are quite short and since they are not used for locomotion can be used for carrying items such as some kinds of nesting materials that are too bulky or unmanageable to go into his pockets. If he happens to be a female and danger threatens, those forelegs are helpful in carrying young ones to safer places. When they are not in use, he keeps them tucked up under his chin almost out of sight. His hindfeet and legs are very long and are powerful saltatorial springs.

One aspect of those hindfeet puzzles me. I can see in it only an evolutionary quirk with no functional meaning. Of his twenty-two species eleven have five-toed hindfeet, ten have four-toed — and one sometimes has five and sometimes four. For a time some classifiers considered this sufficient reason to divide *Dipodomys* into two genera, giving him the four-toed and putting the five-toed into another they labeled *Perodipus* — a procedure that put the four-or-fiver in a sort of limbo, sometimes assigned to one genus, sometimes to the other. Nowadays there is agreement that the difference is not diagnostic, that the one genus is adequate and the digital differences merely an aid in distinguishing species. After all, the fifth toe when present is vestigial, small and on the inside and of little if any use.

But why should some species have eliminated it while others have not? There seems no logic involved because there are instances of both five-toed and four-toed inhabiting the same general areas subject to the same adaptive pressures. *Dipodomys* has been the subject of innumerable scientific studies and I have read many of them and have yet to find even a hint of an explanation of this digital diversity. I have to accept it as just another of the amazing things about him that make him, for me, the most fascinating of all American rodents.

His overall range is similar to that of pocket mouse *Perognathus*, extending from southwestern Canada down through all our western states deep into Mexico. Within it he shows an even stronger preference for arid areas.

Heteromyids are burrow dwellers so naturally *Dipodomys* goes to the family extreme in this. He likes an intricate home of many rooms and galleries on several levels with a variety of doorways and escape hatches. Since he is a professional jumper — and when in a hurry wants to zoom right into his home at the end of a leap — his doorways are comparatively large while his escape hatches are barely body size and usually hidden by being plugged with dirt. Sometimes, when local conditions make this advisable, he will dig drains below his living quarters. He spends much of his spare time repairing and expanding his underground castle.

His small forelegs and forehands, as efficient as those of the others in the family for scrabbling for seeds and stowing these in his pockets, are not the best of digging tools. No matter. His versatile hindfeet take over. While he leans forward on his forelegs, his hindfeet can be brought up beside and past his head to loosen dirt, scrape it backward, and send it flying behind him. A corollary to that kind of action is that he, a professional kicker as well as jumper, can kick forward almost as joltingly as downward and backward.

When some fate overtakes one of him and a burrow is thus unoccupied (no thumps when investigated), another will very likely move in — and continue remodeling and expanding the house-plan. Many a burrow used by successive owners, each adding to the original layout, becomes in time a large mound covering a maze of rooms and tunnels.

Ernest Thompson Seton once excavated such a mound and underlying burrow. He found seventy-five feet of tunnels, ten entrances, seven ordinary rooms, twelve storage rooms, three rooms for dung deposit, with the lowest level a good three feet down.

In such a case the mound and ground beneath may be so honeycombed that when a large something, say a human foot or a horse's hoof, steps on it the result is a cave-in. I have heard complaints involving bad tumbles on that score. My response is that only silly (make that inexperienced) Easterners, human or equine, make that mistake.

Heteromyids are food hoarders, so naturally *Dipodomys* carries this to the family extreme. I know of one instance in which almost fourteen bushels of seeds and dried grasses were uncovered in one burrow system. That was really excessive, probably the work of successive obsessive owners over quite a period with the accumulating stores kept in fairly good condition by the dryness of the region. Between one to two bushels would be a good average for a healthy adult in a good year — which still would be more than twice what he could possibly eat in a year.

Dipodomys is not only interested in quantity of food on hand; he is meticulous in handling it. He wants it dry for better keeping. If it is moist when gathered, he will deposit it in shallow excavations, cover it with loose dirt to hide it while the heat of the sun can still get at it, and when it is well dried transport it to a storeroom. Within his burrow he is apt to stock each kind of food in a separate room like a housewife sorting out groceries and if he has several kinds in one room, he will have them in separate little piles.

The heteromyid social order prescribes solitary adult territorialism so naturally *Dipodomys* is emphatic in following it. Neither he nor she normally tolerates another adult intruding into a home territory. But both of them are under the common imperative of most of the animal kingdom to see to it that another generation arrives on schedule. My impression is that though they obey the command they do not regard obeying as one of life's major pleasures. They indulge in very little courtship, and that very brief, and betray scant indication of romantic rapture. My impression also is that not always but quite often she is the instigator of the proceedings. When she is in estrus, she is more likely to go looking for him than he for her. When she finds him, he recognizes her condition and restrains his usual pugnacity toward an intruder — but may not be inclined to do anything more. She may try to encourage him by being a bit coy, bouncing up to touch noses and back, perhaps leap-frogging over him. If that fails, she may lick his penis until he is sufficiently aroused to do his share of the duty. That done, their excuse for a romance is over and both return to the more usual business of gathering food and guarding their individual territories.

Thereafter, for the brief period necessary, the female takes motherhood seriously. When her time approaches, she brings in materials to make one of her rooms into a snug nest. She can produce up to three litters a year, averaging three to four young each. They are born tiny, almost hairless, pinkish in color, quite wrinkled, with eyes and ears closed and hindlegs and tails much shorter proportionately than later they will be. In about two weeks they will be furred and their eyes and ears will open. In about two more they will be quite active and the territorial imperative will already be operating in them as they practice establishing small territories against each other within or close to the burrow. In about another two weeks, pushed to it by their mother if they are reluctant, they will be on their way looking for real territories to claim as their own.

Heteromyids thrive on dry food (except for some greenery to encourage mating) and need no water. In the wild they are free of any worry about being within reach of any waterhole. Previous notions that they obtained moisture by tunneling down to ground-water, by eating succulent greenery, or by collecting dew at night have long since been discarded. Kept as pets, they can live out their lives on an intake of nothing but dry seeds. Except under contrived laboratory conditions any water provided will not be used — in fact will be avoided. Heteromyids are annoyed if any part of them ever gets wet. They need no water.

I am playing with words. Of course they need water. The point is they do not need to drink any to have some. They have solved the water problem of arid-land living by reducing their requirements, by conserving what moisture they do get from their food — and by making what they need. This too is a combination of techniques of which they are not the sole inventors. Convergent evolution applies again. Other small dwellers in arid lands in other parts of the world have evolved similar techniques. But I can claim that our heteromyids have done a very good job with them.

I do not know whether *Dipodomys,* the kangaroo rat, extremist of the family in many respects, is the most efficient in water conservation and manufacture. I suspect that he is, though I may be misled by the fact that so many careful studies have been made of him that more is known about his use of the family methods than about that of any of the others. Certainly he does seem to be particularly efficient at living in hot dry regions. He can and does live quite successfully (and apparently happily) in the depths of Death Valley where it would be impossible for him to obtain fresh water. Under such conditions he is faced with a formidable challenge. He has to meet not only the problem of lack of available water but also the companion problem of the heat of the sun. And normally, for most of us mammals, a major defense against heat is the use of moisture.

Curiously enough *Dipodomys* has little tolerance for heat. If the temperature about him rises to 100 degrees Fahrenheit, he is in serious trouble. If it rises higher he will die within a matter of hours. Yet surface temperatures in desert areas regularly rise much higher. He avoids them by digging his deep burrows and resting quietly in a lower chamber during the daytime, becoming active and coming out only at night when ground and air are cooler and humidity is higher.

He has to follow such a schedule because he has no sweat glands, has foregone the use of moisture evaporation for cooling — tries even to avoid the carnivore method of cooling by evaporation from tongue and nasal passages. This is one means of conserving his internal water supply, but it can sometimes put him in a serious predicament. Suppose a predator (certain snakes consider him a dietary delicacy) invades his burrow in the daytime and he is forced to take refuge in one of his escape tunnels. If he pushes on out into the open, he is exposed to another enemy, the heat of the sun. If he stays in the escape tunnel, having plugged the inner opening in hope of not being detected, he is still close to the surface and the daytime heat. In such circumstances he resorts to an excessive flow of saliva, dampening the forepart of his body in a desperate attempt at that kind of emergency evaporative cooling. This imposes a drain on his moisture resources, but if the emergency does not last too long, he will survive it. I like to think that usually he does.

He has other means of water conservation. All of us lung-equipped breathing animals regularly lose much moisture in the form of that absorbed from our lungs by the air we breathe out. I make use of that simple fact by huffing on the lenses of my glasses so that moisture will condense on them when I want to clean them. If *Dipodomys* wore glasses, he would have difficulty doing the same. He has found a way to recapture much of the moisture he might lose when he breathes out. His nasal temperature is kept lower than his lung temperature and his nasal passages are quite long and labyrinthine. On its way out the air is cooled enough to leave behind some of the moisture it has taken from his lungs and this is reabsorbed into his system.

Again, he has found a way to reduce to a negligible point the loss we other mammals spendthrift of water suffer without even thinking about it every time we void our urine. His kidneys are at least five times as powerful as ours. They do not use as much water as ours do in performing their function. More important, they and his urinary bladder can reabsorb much of what water is used. His urine as voided contains very little moisture, is highly concentrated — more so even than that of the camel, who was once thought to hold the record. He could drink sea water with easy impunity — except for the laxative effect of some of the salts it contains. Chemists have pointed out that he must have some means of avoiding crystalization of that superconcentrated urine in his bladder, but as far as I have found out to date that remains his own secret.

He applies that reabsorption technique to his feces. As voided they are hard dry pellets. Even so, like the lagomorphs, he may re-ingest them, perhaps for a second chance at extracting any moisture (or food value) missed the first time through his digestive tract, but more likely in order to obtain vitamins synthesized by bacteria in his large intestine.

So *Dipodomys* does not need as much water in relation to his size as do most of the rest of us. But where does he obtain what he does need? Some from the food he eats, yes, since some moisture is always present even in what appears to be completely dry food. Yet that is only a small portion of what he requires. He gets the rest of it in the same form that all of us get some of our water — "metabolic water," that formed within us by oxidation of hydrogen in the foods we eat.

That explains his preference for seeds and certain grasses. They have a high percentage of the kind of carbohydrates he uses in creating water. Which in turn, since this water is the result of metabolic activity, explains why there is a direct relationship between how active he is and his food intake. If he is given only food with a high moisture content (not his personal choice), he becomes rela-tively inactive. He is obtaining more water than he normally wants from his food. If he is given dry food only, he becomes more active, sometimes hyperactive, thereby increasing the metabolic water-mak-ing process within him to match his need. When he is bustling about in the wild, then, he is not only adding to his food supply, he is adding to his water supply too.

But it is not quite as simple a proposition as that. I am one of those who used to boast about him that he had licked the desert problem by developing a special ability at manufacturing water. I know now that he has not. He is not much better, perhaps no better at all, at synthesizing water by metabolic action than we other mammals.

All of us add to our water supply when we are active. But at the same time we increase our water loss through heavier and faster breathing and increased evaporation through our skin with its sweat glands. Our loss outruns our gain. Because of his conservation techniques, *Dipodomys* achieves a reversal, a slight margin of gain.

My boast needs revision. This little fellow who looks like a caricature of the rat he is not has licked one of the toughest environmental problems any living creature can confront by even more ingenious special techniques than the common one of synthesizing water.

I have stated that I do not know whether *Dipodomys* is the family extremist, the most efficient, in solving the water problem of arid land living. I have no doubt at all that he is the extremist in saltatorial gymnastics.

This too has a direct connection with his choice of habitats. Arid regions, even semiarid, are chiefly open country with scant vegetation and what there is quite scattered with bare spaces between. When food-gathering, *Dipodomys* has to operate where he is often very visible and thus exposed, almost overexposed, to predation. For survival in adequate numbers he needs space-covering speed. His long strong hindfeet and legs give it to him. In proportion to his size he is a prodigious jumper. His gait when merely ambling along is jerky and seemingly awkward, a series of short little hops. When he shifts into high gear (he can do that instantaneously) he becomes something amazing, appears to be jet propelled. Even his smallest species can zoom along in five-to-six foot leaps, his largest in ten-foot. He can be from here to there faster than other small rodents who hold to the conventional four-footed ground-hugging style of locomotion.

But speed is not enough. Most of his predators are much larger and this gives them greater straightaway speed than he can muster. And winged predators can overtake even a stretched-out jack rabbit. His major defense is his wondrous ability at evasive tactics. Here his long tufted tail is of vital importance. It is his balancing rod, his steering rudder.

He can leap straight up or forward or backward or sideways at any angle. He can change course in midair. Seeing him in swift escape action can give the impression that when he takes off in a full leap even he himself does not know where he is going to land. He will decide while soaring. He is completely unpredictable in flight. So much so that an experienced owl, say, having missed him on a first strike, will not bother to try to follow up with a second.

There is something uncanny about his tactics, something involved beyond speed and agility and instant reaction. His eyesight is not particularly good except perhaps for the close-up work of picking seeds out of sand and for that his nose may be the better guide. He is out and about primarily only at night, when visibility is poor anyway. As a matter of fact, the darker the night the better he likes it. In time of full moon he may not come out at all. The way he behaves when he is out suggests that he deliberately makes little use of what protective covering is available in his territory. He moves freely about in the open. Is he depending on the dark to hide him? Definitely not. He has deadly enemies to whom darkness is no handicap in their hunting.

There he is moving slowly around on some dark night, stopping to stuff his pockets with seeds. He is as adept at this as his pocket mice relatives, but he does not do it with the same swift intensity and hurried dashes back to his burrow and he ventures much farther afield. He appears to be too intent on his work to be paying much attention to his surroundings. He is coming closer to a coiled rattlesnake who is already aware of his presence. Does he know that danger threatens? Apparently and quite possibly and even probably he does not. Suddenly the snake strikes — and *Dipodomys* simply is not there. In the last tiny fraction of a second while the snake is in motion he has zoomed up and away.

Again he is intent on his pocket stuffing. An owl has spotted him from some distance away and is floating softly toward him. Does he know it is approaching? Apparently and quite possibly and even probably he does not. He has not looked up from his work. Swiftly the owl drops, talons ready — and again *Dipodomys* is not there. In the last tiny fraction of a second he has zoomed off at an angle and away. If he had taken off sooner (as also with the rattlesnake) he might have given the enemy a chance to change course and perhaps hit him in midair or as he came down. He has waited until each was in the midst of its strike, committed to the precise spot he was expected to be occupying, before triggering his vanishing act. The snake's fangs have missed; the owl's talons have grasped nothing but sand. And even now *Dipodomys* does not move very far away and displays no particular alarm or haste.

How does he do it, sense unseen danger striking swiftly and seemingly silently toward him? That is a survival technique best discussed in connection with the last member of the family.

Merriam's *Microdipodops,* tiny two-foot the kangaroo mouse or dwarf kangaroo rat, has a quite limited range, primarily the Great Basin of Nevada with a few outposts in the edges of the surrounding states of Oregon and California and Utah. His species total also is limited, just two. He looks like a small version of *Dipodomys* with two noticeable differences. One is that his tail has no tip-tuft or white stripes and is bigger in diameter in the middle than at either end. Sometimes that middle is bigger than at other times. It probably is a storage place for fat and especially useful as a source of extra energy for a female during pregnancy. The other difference is that his hindfeet have fringes of stiff hairs which help give him traction in the fine loose sandy soil of the areas he inhabits.

He has all of the heteromyid habits and characteristics in his own degrees. He has all of the heteromyid survival techniques, very definitely including that uncanny ability of *Dipodomys* to sense unseen danger.

Some pages back I wrote that heteromyids have bullae, auditory chambers of the middle ear, proportionately larger than those of other rodents, probably than those of any other mammals. I add now that their bullae are progressively larger from genus to genus in the sequence in which I have been discussing the genera, from pocket mouse to kangaroo rat to kangaroo mouse. Little *Microdipodops* holds the record. His hearing and response mechanism comes close to being a match, though in a different way, for that of the echo-recording bats.

Much more is involved, of course, than the bullae alone. I use them as a symbol of the whole mechanism because they are a key factor in it and because they are the most noticeable factor, accounting for the big-headedness of the family — which again is progressive from genus to genus with *Microdipodops* holding the record. The whole mechanism is wonderfully sensitive for catching and transmitting the slightest of sounds — and triggering the appropriate muscular response. The bullae are what add magnification of those sounds so that they can be correctly interpreted — which is very important for *Microdipodops* (as for *Dipodomys*) because he lives where there is very little vegetation and of necessity he has to be often out in the open and because his enemies offers sounds only of very slight magnitude and of very low frequencies.

Experiments with delicate instruments have shown that an owl dropping down for a strike produces a very slight whirring sound with frequencies up to only 1200 cps (cycles per second). Just before striking, a rattlesnake produces a very faint rustle, probably from scales shifting a bit as muscles tense, which increases a trifle with the motion of the strike itself but has frequencies only up to about 2000 cps. Most small rodents have little or no auditory sensitivity to sounds below 3000 cps. We humans, of course, have none at all. *Microdipodops* has his greatest sensitivity to sounds between 1000 and 3000 cps.

Out there in the desert spaces it happens often. An owl swoops down, a rattlesnake strikes, and tiny *Microdipodops* is the target. To a human observer it would be an absolutely silent drama. Not to *Microdipodops*. He hears the owl dropping toward him. He hears the rattlesnake striking. And he does not panic. He has the quiet courage to wait to the last thin margin of safety that is the true margin of safety before making his own move.

Always, when nowadays I consider and contemplate of any of my fellow mammals, a feeling of wonder and of what can only be called appreciative awe creeps through me at the variety and ingenuity of their answers to the common problems all of us face in living on this planet Earth. I try to be objective, to indulge in no favoritism — at least only a minimum of it. But I must confess to a special fondness for the heteromyids, the "different mice." Perhaps that is because they are the ones who really started me on my erratic and stumbling wanderings through the maze of mammalian creation.

The time was I too merely took them for granted as nothing more than variants of the ubiquitous mice and rats who happened to share my liking for life in the Southwest. So stating I do not mean to put down mice and rats, whom I now know are amazingly various and versatile in their own multiple ways. But I had been aware of mice and rats all my life and in my ignorance thought them all much alike and simply scrubby little creatures worth attention only as nuisances. About the only aspect of these southwestern variants I had noted was that they were not nuisances, had no desire to invade human households or barns and chickenhouses and such. Then my wife and I went on a camping trip into what is now Canyonlands National Monument.

I was supposed to be absorbing impressions for an article about that stupendous Standing Up Country, about the tremendous sculptured topography which nowadays regularly astounds tourists. I know very well that what I wrote remains only a thin pallid reflection of the awesome geographical reality. But what I remember best is none of that. It is a small living thing, *Perognathus,* the pocket mouse.

The first evening we were camped on a lookout point, the vast gorge upon gorge of the Green River dropping away below with some of the tall Standing Rocks down there on the second level. The prospect dwarfed a mere human into a minuscule item ashamed of being so infinitesimally tiny in the immensity. And there, quite at home on that promontory, untroubled by his own even tinier tininess, the joy of existence tingling in him, was *Perognathus.*

He had not yet learned to distrust humans. He regarded us as welcome and interesting visitors. He appeared in the early evening and thereafter skipped about, observing and investigating everything we did and everything we had. He thought some crumbs of our food were worth being tucked into his pockets. He sat briefly on one of my boot toes and looked up at me as if he wished we could carry on a conversation. He was still skipping about, right over our bedrolls too, when we finally went to sleep.

I saw him then as I should have seen him long before. A neighbor, a friend, a companion spark of life in the vastness of the universe. Perhaps he was really only a mouse, but in the closeness of that evening's companionship I could see he was definitely a different mouse. And so, after that trip was over, I began to learn what I could about him. A fascinating trail to follow. And of course it led me on to *Dipodomys,* the truly saltatorial kangaroo rat, and at last to *Microdipodops,* that compact little edition of heteromyidism who sums in his tiny self all of the heteromyid adaptations and techniques and simple courage for life in my Southwest.

"Sir," said the man named Buckthorne in Washington Irving's *Tales of a Traveler,* "there are homilies in nature's work worth all the wisdom of the schools, if we could but read them rightly." Just that and nothing more and nothing less is what I have been trying to do through all my considering and contemplating of my fellow mammals — trying to read those homilies rightly.